BREAKING
the RIDDLE

BREAKING
the RIDDLE

**Time vs Space, and the Universe
beyond the Standard Model**

Christos Tsikoudas

ISBN 978-618-00-3523-0

Cover photo: andrey_l/Shutterstock.com

In memory of Minas Vavakos,
who guided me on how to pursue rational thought
at the age of young.

Contents

Overview

This publication provides a first-of-its-kind approach on the essence of time and space and their role on how the universe functions at various scales. It introduces concepts that lie beyond the context of the Standard Model of physics, and get to effectively address long-standing open issues in areas ranging from cosmology and astrophysics, to elementary particle physics. Particular attention has been given in presenting these concepts in a way that allows comprehension from readers of various levels of knowledge, providing a strong intuitive grasp on how the cosmos operates.

The presence of mass is known to curve spacetime, and as a result time passes slower near a celestial object than it does far away from it. But, does time keep moving faster and faster, or reach a flat value, the further away toward the very distant outer space? It turns out that neither of these conditions hold, as beyond certain range far away from our celestial neighborhood, time appears to start moving slower and slower toward the very distant outer space, as part of a particular loop-gravity effect. That opposite inclination of curvature is explained to gravitationally force distant celestial matter further away, for the same reason it attracts matter toward black holes. That constitutes the cause of the observed accelerated expansion of the universe. In such way, the curvature of spacetime at the cosmic scale is explained to feature a characteristic (radial-like) waviness, of which, the observed universe concerns a positive portion (antinode). The inclination at either side of this waviness designates where, and how strongly, matter is attracted toward. The inclination of curvature toward the surface of black holes attracts celestial objects toward that side, while the inclination towards the very distant outer space attracts celestial matter radially away toward what we consider classical infinity, where the metrics of space and time eventually turn negative, with particular implications. This model is analyzed to explain currently unresolved astronomical observations.

Despite the strong interconnection between space and time, the waviness of the SPACE metric is explained to be subject to a characteristic phase shift with respect to the waviness of the TIME metric, which may remind of the phase difference between the Voltage and Current oscillations in electric circuits. The underlying symmetry between these two cases, is explained to be non-coincidental. This phase shift between the curvatures of the space metric and the time metric is explained to refer to the Higgs field, that permeates spacetime. It is also discussed that the value of this shift in terms of degrees of angle matches the Weinberg angle of the electroweak interaction.

This concept allows to describe in quantitative terms how the Higgs field mechanism applies in causing the inertial behavior of matter, in symmetry to as an inductive impedance reacts the operation of electric engines. This symmetry to the theory of electronics also extends to address how a particular waviness of the metrics of space and time at the cosmic scale may explain the observed abundance of matter over antimatter in the universe.

Moreover, effects of fluid mechanics like vortex flow and the Bernoulli pressure are shown to exhibit an analogous symmetry to reaction effects of electronics relating to induction and capacitance, which reveals a specific orthogonality between mechanics and electronics. In relation to that, much light is shed in understanding how forces physically apply in both mechanics as well as in electronics, while new grounds are introduced in distinguishing between action forces (like the gravitational, or the electric) and reaction potentials (like the inertial, or the electromotive), which were previously lacking adequate understanding.

An additional key ingredient of the new theory is that it provides a new way to physically interpret equation's imaginary terms, through a process of negative-reciprocation of parameters. This process is shown to concern and explain specific non-classical attributes, like the nature of the Planck constant, and in such way resolve the origin of quantized behavior. It also reveals that certain equations of electronics already incorporate a negative reciprocation of parameters into them without that been previously realized. By extent, light is shed on particulars concerning the difference between bosons' and fermions' essence. In addition, an association is made possible between quantum nature and effects of relativity concerning a particle's mass and momentum.

The presentation of the new theory starts with a phenomenological approach, explaining how the metrics of time and space set up an underlying structure, whereupon elementary particles reach form and interact. That is followed by certain guiding assumptions, as well as quantitative approximations, which bring up a new viewpoint that drastically supplements the understanding of nature's interactions.

The whole model has been developed through private research that lasted over two decades of time, and went through countless iterations. Even though new development has never stopped arising per iteration, I have decided to publish material already available, so as to open up the way for further discovery and improvement by the scientific community, along the road that leads toward a complete Theory of Everything.

This new theory additionally opens up a large window for future technological innovation, across numerous areas of science. The author has already carried out significant core work (not included in this publication) on the scientific domains of remote propulsion by electronic means, control over resonant optoelectronic interactions, and ultrafast telecommunications.

The prospects for the future seem to be mostly promising.

Christos Tsikoudas

1
Waviness in Spacetime

According to the theory of relativity, the existence of matter curves spacetime. One demonstration of this feat on earth, is that the time metric reads lower value at lower altitude, than it does at higher altitude (i.e. a clock at sea level ticks slightly slower than an identical clock on a hill). That feat is known to reach a minimum at cosmic regions like the surface of a black hole, where time is considered to reach a halt. In an attempt to describe schematically such curvature of the time metric and extrapolate it over the wider cosmic scale, the graphical representation of figure 1-1 has been prepared. In this figure, the "Time Curvature" schematically represents in simplified form "how fast the time is progressing" in various regions of the cosmos. Or, in more technical words, the amplitude of the Time Curvature at each point along the axis of the figure represents the magnitude of the vector of time in that region.

Let us imagine that the horizontal axis of this figure corresponds to an immensely large hypothetical path, coming from as far as the very distant outer space, passing from the neighborhood in our solar system, and diving into the black hole at the center of our galaxy. Regarding the Time Curvature, if we read it in the direction from left to right of figure 1-1, starting from point a', this point is assumed to correspond to a place at distant outer space which is far away from any galaxy so the scarcity of matter at that place (and the corresponding negligible curvature of spacetime) gets the Time Curvature to read a large value, meaning that, at that cosmic region time progresses fast. At the right side of it, point S corresponds to the region of our cosmic neighborhood (where "S" stands for our "solar" system), where the Time Curvature has a somewhat lower value, meaning that time is passing slightly slower around our celestial neighborhood. Further to the right, the Time Curvature is shown to reach the value of zero as it reaches the surface of our galaxy's black hole which identifies to point b'. Beyond point b' to the right (inside the black hole) the time metric is speculated to take negative values.

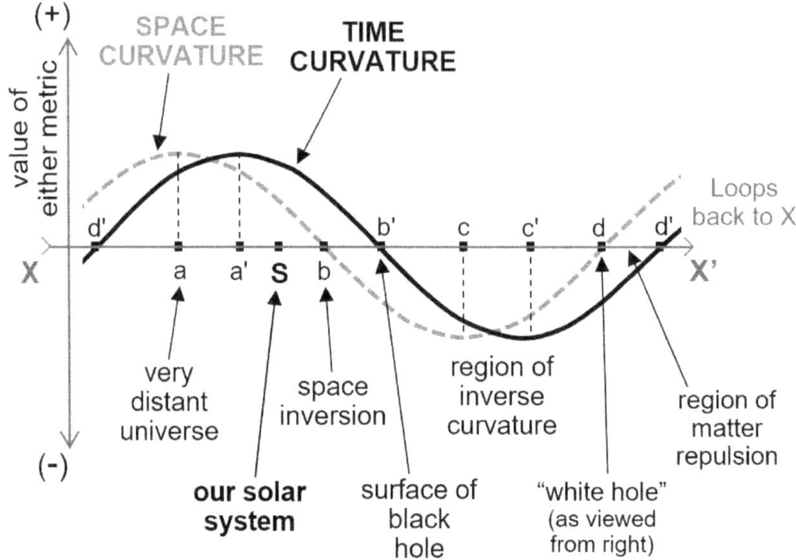

FIGURE 1-1: Schematic representation of the waviness of the time and the space metrics (normalized) over the cosmic scale. (High amplitude means time passes faster, or space is more stretched).

In the other direction, reading to the left of point a′ we postulate that the time metric takes decreasing values again, which translates to the time progressing slower the further away toward very distant outer space (like in mirror to what was the case for the metric of time in approaching our galaxy's black hole). This opposite slope heading toward the very distant outer space, signifies gravitational attraction of celestial objects outward (toward what we consider infinity, in the direction from point a′ toward point d′). This feat is responsible for the observed **accelerated expansion** of the universe. The functioning of the gravitational force at that distant region is no different than the gravitational action at our cosmic neighborhood, it is just that the opposite inclination of slope of the curvature in that region corresponds to the gravitational acceleration that points in the opposite direction (toward the very distant outer space). Actually, the further away to the left from point a′ toward point d′, the stronger the steepness of the Time curve, thus the stronger the attraction of celestial objects in the outward direction (seemingly toward infinity), hence the observed accelerated expansion of the universe.

We shall now move a step further and make another assumption, which may seem arbitrary for the moment, but will acquire meaning and justification as we move on in this theory. That is, we shall assume that the right end of the horizontal axis of figure 1-1 loops back to re-attach to the left end of the same axis, in forming a closed loop. In other words, the horizontal axis of figure 1-1 can be understood to correspond to an immensely large conceptual loop (a sort of a hypothetical immense "string") which has been graphically "opened-up" into a straight line (XX') in favor of graphical representation. According to this new assumption, the Time curve at the left of point a' eventually catches up to the right side of the figure 1-1 (this is why the same point d' is shown at both ends of the axis, since the axis forms a loop). That nodal point d' actually resembles a form of event-horizon (lying at the very distant outer space, within the positive space metric), where the conventional universe reaches a limit, as the Time metric reaches a zero value again.

Let us now consider the case of a similar characteristic of waviness applying to the curvature of the Space metric as well. One reason for that to be happening, is that the curvatures of time and space are tightly correlated to each other. In fact, speaking of space, in certain cosmic regions like near black holes, space is more compactified (in association to the spacetime curvature prevailing at that region), while in certain other regions, like in far outer space, it is spread over "vast vacantness". That suggests that the space metric should exhibit a similar wave-like curvature (in terms of space contraction and space stretching), just like the time metric was presented above to do. The Space Curvature of figure 1-1 attempts to schematically represent the corresponding shape of such space curvature at the cosmic scale, and this curve has been purposely shown to be a little shifted to the left of the Time curve, for a reason to be figured later. The high amplitude of the Space curve in the neighborhood of point "a" is meant to represent the stretched-out metric that prevails at far outer space, in full accord to the celestial vastness of that cosmic region. Correspondingly, the low amplitude of the curve in the neighborhood of point b is meant to reflect the stretched-in (contracted) space prevailing near a black hole, in accord to the small celestial range a black hole occupies (despite the huge contents accumulated in that celestial region). We shall also figure that there also exist contracted space regions at the very distant outer universe, associating to point d (as "viewed" from its right). These regions seem to correspond to distant celestial objects of unconventional behaviors, toward which space seems to contract as well, and we shall later figure that this seems to refer to celestial objects like the quasars.

Along these lines, the schematic depiction of the full Space Curvature in figure 1-1 should be understood to incorporate a graphical "normalization", which allows to graphically fit, and represent, both the celestial regions of seemingly infinite size (at distant outer space), as well as celestial regions that appear small in range, but contain lots of contents (like space near a black hole). This graphical representation also makes it easier to extrapolate in describing what may hold in regions of unconventional space of negative metric which we shall consider later.

The depiction of the whole spacetime waviness in figure 1-1 is obviously an oversimplification, since in the actual universe there exist innumerable black holes and immeasurable extend of distant space, therefore there exist infinite corresponding hypothetical pathways through the cosmos. In this sense, the presence of so many different celestial formations in the universe comprises a whole "sea of waviness", which may have local positive and negative antinodes of unequal sizes. Therefore, figure 1-1 is only meant to describe in simplified form the concept of **waviness** as well as the "**loop-ness**" (since the axis of figure 1-1 forms a loop) of the time and space metrics in nature.

Striking and exciting as it may seem, same as some centuries ago humans were called to conceive that earth isn't flat, but is actually a sphere, we are now just as like called to realize that space at the cosmic scale does not extend infinitely in a "flat-like" way, but it actually follows a wave-like formation, which at some parts takes negative values, and eventually forms an immense loop. And time alike, does not "pass" at a classical rate we think of, throughout the cosmos. It also follows a characteristic waviness as well as loop-ness, and in some cosmic regions time progresses negatively, where interactions still fully obey causality, under a different balance where different particles reach stable form, as we shall explore below.

2
Phase Shift between Space & Time

In the depiction of figure 1-1 the space and time curvatures are shown to have a phase difference between them. The existence of such shift allows for providing an explanation to a series of physical phenomena, just like a phase shift between the voltage and current oscillations explains numerous effects in electronics. It is therefore critical to seek for evidence for the existence of such a shift.

It turns out that at the cosmic scale there does exist a quantitative prediction that matches this assertion. It concers quantitative findings, according to which there exists a particular surface outside our galaxy's black hole, where, if an object orbits outside of this particular surface (on "our" side of it), it feels a centrifugal force heading outward, while if it orbits inside (at the other side) of this surface it feels a centrifugal force heading inward. Or, light rays at the other side of this surface are predicted to bend in the opposite direction (outward, instead of inward toward the center of the black hole), and even further, angular stresses are likewise predicted to transfer momentum in the opposite direction.[1] We shall here correspond this particular surface (actually, a form of "horizon") to point b of figure 1-1, coinciding to a node of the SPACE metric's curvature. The black hole's event horizon, in the other hand, coincides to a node of the TIME metric, lying at point b' of figure 1-1. This suggests that the TIME metric and SPACE metric curvatures have a phase shift with respect to each other, of extend analogous to the segment b-b'. Such a shift should not be only local, instead it should extend throughout the loop-like axis XX' of figure 1-1.

It is of significance that this shift between the space and time curvatures is born for similar reasons, and shares similar characteristics, to the ones of the phase shift developing between the voltage and current oscillations in an electric circuit. The symmetry between these two cases is not coincidental. In the electric case, the voltage-current phase shift depends on the "load" that circuit elements (coils or capacitors) provide to the system in relation to the fields and corresponding forces generated. For example, in an industrial facility operating electric motors, the oscillation of current gets to lag the oscillation of voltage (unless specific balancing measures are taken to counteract it), and this associates to effects of reaction (e.g. inductive impedance) which involve the electromotive force. Through a symmetric process, the shift between the Space and Time curvatures exists in relation to a "load", which concerns the energy trapped in acceleration of all matter within the universe, from elementary particles, to whole celestial bodies that these particles comprise. And this phase shift between space and time, in turn, will be shown to be responsible for generating the inertial behavior of matter (meaning the impedance to a change in the motional status of matter), through a very symmetric process to the one of the inductive impedance in electronics. In this sense, the phase shift between the metrics of space and time constitutes one of the key reasons of the of the symmetry breaking in particle physics, and lies behind the Higgs field mechanism which "dresses" matter with mass. This phase shift extends throughout the cosmic scale, and actually directly refers to the Higgs field that permeates spacetime. In fact, the reason that this field interacts in the form of a boson particle is completely distinct, and shall be explained later to arise for similar reason that electromagnetic fields oscillations interact with particles in the form of photons.

Furthermore, this particular phase shift (as depicted in figure 1-1) will later be discussed to match and identify to the weak mixing angle of the electroweak interaction (the Weinberg angle). The existence of this shift allows nature to provide reaction to change, by getting a potion of energy that is being spent to be consumed withoutout delivering work. This becomes possible through the involvement of imaginary terms, as we shall consider throughout the next sections. The particular value of this shift (approx. 30 degrees) also appears to be the reason of ½ of the value of the Planck constant in an electron's ground state, as well as of the value ½ of the electron's intrinsic spin as we shall also consider.

Moreover, similarly to as in electronics the value of inductance or capacitance affect the frequency and damping strength of an oscillation, the phase shift between the space and time curvatures (which associates to inertial impedance) along with pressure-related differentials (to be associated with a capacitive type of behavior) similarly affect the frequency (rapidness of occurrence) and damping rate of mechanical effects at various scales, from small, to large celestial.

In a point of attention, any symmetry between effects involving the phase-shift in the curvatures of Time & Space, with effects involving the phase-shift in oscillations of Voltage & Current, is not by coincidence. It shall be shown to stem from a fundamental symmetry between the gravitational and electric interactions, which involves a specific swapping between parameters of space and time, as shall be described over the next sections. In one example of such swapping, consider that in electronics a phase shift typically develops in TIME (for example in an *L-C* circuit the alternation of current gets to lag-or-lead in time the alternation of voltage), while in mechanics the phase shift typically develops in SPACE (like in the depiction of figure 1-1).

Furthermore, it is of particular significance that the phase shift illustrated in figure 1-1 allows for four combinations of positive and negative values of the space and time metrics. Namely, Positive Space & Positive Time (between points d′ and b of figure 1), Positive Space & Negative Time (between points d and d′), Negative Space and Positive Time (between points b and b′), Negative Space and Negative Time (between points b′ and d). That allows for the existence of four distinct types of cosmic regions, each of which will be explained to constitute stable habitat for different kinds of elementary particles, with symmetric variants of the known interactions applying in each of these cosmic regions.

3

Negative Spacetime Metrics

According to the theory of relativity, if a traveler moves at a high relativistic velocity, his/her wristwatch tics slower with respect to the clock of someone at rest on earth, due to the effect of time-dilation. In theory, if the velocity of light were reached, time would reach a halt. A question of interest is that, supposedly a fictitious traveler -made up of some exotic substance- could accelerate to faster than the speed of light, then would they experience time move backwards? To explore the case, let us introduce some concepts which come across what the negative time may imply.

3a. What Negative Time Stands For

First of all, it should be made clear that if time moves in the negative direction, this has absolutely nothing to do with things moving backwards like in a video run backward. Let us consider a hypothetical region in space where the time metric has a negative value, either a tiny local spot, or a large region like cosmic region $d - d'$ of figure 1-1 (where the space metric is still positive). A place where time passes in the negative direction should rather be understood as a place involving an opposite form of energy, and as we shall figure ahead, in a case like that different elementarty particles reach stable form, so a different balance is reached. For an elementary particle that maintains stable form and interacts in that region, the cause-and-effect conditions apply regularly, yet this is done solely with respect to the negative direction of time (without the "positive" time being involved in parallel for purposes of observational measurement).

If we hypothesize a time-travel experiment, having a capsule to travel in the negative direction of time and record what happens in its surroundings, this runs into serious obstacles. Aside energy requirements, another major obstacle seems to be that we may not manufacture a capsule made of some exotic substance that could survive living in negative time. This is so, since -in relation to material to follow- the conditions of negative time appear to favor the stability of antimatter, while concurrently, conventional matter would rather be unstable. In that sense, conventional matter should be subject to disintegration, via interactions including annihilation events, happening for same reason that positrons are not stable in the classical world. In addition to that, any "video recording" processes that a hypothetical capsule would shoot, would record effects that develop on the spot given the negative energy conditions, along the involvement of exotic particles (actually, antiparticles) that survive in negative time. In this sense, the video recorder would simply record its own history evolving regularly along the involvement of local anti-particle inhabitants, interacting physically, chemically, and biologically with each other. That has nothing to do with "viewing" interactions of conventional matter which took place in the past. Therefore, the notion of a capsule that gets to time-travel to the past and come back to reveal to us what happened in "our" past, where "our" past concerns the molecules that were composing us in interaction to all their environmental influences at any level (including the molecules of air around us, rain that dropped on us, interactins with micro-organisms inside our body, etc.) is completely erroneous.

As for particles (and larger objects) that already reside in a cosmic region where the time metric has negative values, these particles' interactions definitely obey causality, but through a different balance, where symmetric interactions in opposite energy should prevail, depending on the values of both the time and space metrics at the specific cosmic spot. For instance, in region $b' - d$ where both the Time and the Space metrics are negative, objects would not be able to consist of conventional atoms & molecules in chemical and biological balance as we know of. It would be more realistic to consider that particles of a negative form of matter (to be specified latter) would survive and possibly engage in similar atomic-like formations, and these would become involved in interactions which may be approximated quantitatively through imaginary numbers and terms.

A more conventional region is one where the time metric is negative but the space metric remains positive, like the region $d - d'$ of figure 1-1. According to theory that will follow in the next section, this cosmic region seems to provide favorable habitat for antiparticles (like the positron), as these appear to comply to conditions of negative time in accordance to the Feynman-Stuecklberg interpretation[2],[3], and should consequently survive for long lifetimes in that place. While instead, conventional particles like free electrons should have short lifetimes in that region, as they would rather engage in annihilation events. In any case, supposedly a hypothetical traveler visited such region, he/she would definitely *not* find that place favorable in terms of living conditions, having things simply evolve "backwards".

A question of interest is, what's the closest we (humans) may experience in terms of negative flow of time. Surprisingly enough, other than the particular association between negative time and antimatter (since positrons resemble electrons in negative direction of time), effects involving negative flow or backward shift in time, appear to be taking place extensively all around us, in effects of magnetic action (later to be associated to curvature in the negative-reciprocal of time) or induction (like in the case of an inductive phase lag).

3b. What Negative Space Stands For

Just like it holds with negative time and the curvature of its corresponding metric, it similarly applies for negative SPACE. Space of the positive metric refers to what we are accustomed to measure as per our classical perception, while that may also be subject to effects of relativity (contraction or stretching) depending on the velocity of an observer travelling in it. Space of the negative metric, instead, may be understood to refer to what a hypothetical observer -or rather, an exotic particle- traveling much faster than the speed of light would perceive. It constitutes an unfamiliar notion which does *not* concern "space in the opposite direction". In relation to what we shall consider ahead, it concerns a place where only specific exotic particles of negative matter (to be specified ahead) may reach stable form.

As mentioned earlier, the space metric becomes negative at the right side of point b of figure 1-1, where that includes the region b - b' outside a black hole, where certain forces are quantitatively predicted to point in opposite direction as previously referred to.[1] Likewise, space becomes negative at the left of point d. Approaching a celestial region of negative space may be considered plausible since (i) space is continuous, and (ii) a region of positive space attaches (is adjacent) to a region of negative space. Such continuity may actually allow energy to conditionally shift from one side to the other during interactions. However, for conventional matter to move through the nodal surface of point d and reach inside the negative space metric would rather encounter serious practical obstacles, since conditions seem to be adverse for classical matter to stay in existence in the negative side. A cosmic region where the space metric has negative values appears to provide stable habitat for particles of negative energy, and as we shall soon consider, this should concern neutrinos and antineutrinos. (We purposely limit the reference to leptons only, since for composite particles the case is more intricate due to specific properties of quarks which shall be explored in section 10).

Along these lines, the cosmic region b′ – d of figure 1-1, and more particularly its outer border d as we "view" it from its right side, may be understood to correspond to the surface of celestial objects of certain unconventional characteristics, like emitting very intense radiation or seeming to rotate at superluminal velocity, **quasars** being one of this type[4],[5]. Actually, due to the opposite sign of the time-curvature's inclination outside such celestial bodies (from the nodal surface d until the nodal surface d′ of figure 1-1), these objects seem to operate as "white holes" (the opposite of black holes) repelling away conventional positive matter and bending away photons, which contributes to the high luminosity of these celestial objects. In relation to concepts to follow, it is also likely that the region b′ – d may produce Z^0 bosons behaving much like photons do in the classical part of the universe.

Other than its prevalence in these remote cosmic regions, negative space seems to also engage quite extensively all around us, as it literally exists "behind" every single point of conventional space. This allows negative space (or a displacement shift backwards in space) to interact with the imaginary part of particles' wave functions, and engage in effects of mechanical impedance, like inertia or dragging, which we shall consider later, thanks to the relation i^2=-1. Furthermore, in section 5 we shall also describe a critical association between the curvature of negative-reciprocal of space and the electric field.

3c. Loop-ness involving Space and/or Time

While the space and time metrics may also take negative values as depicted in figure 1-1, this allows for the curvature of the metrics of time and space at the cosmic scale to exhibit a form a waviness. On top of that, this type of waviness may also be loop-like, meaning that the axis XX' may actually form an immense loop. Such a model of curvature sets no requirement for existence of singularities inside black holes, neither it requires a Big Bang in order to explain cosmic evolution, or the accelerated expansion of the universe. Alongside further particulars to be presented ahead concerning the balance of energy in terms of all particles and interactions throughout the whole cosmic loop illustrated in figure 1-1, the universe may be understood to constitute a self-sustained total system, whose total energy sums up to null.

Furthermore, the whole model of waviness of the metrics of space and time at the cosmic scale described in figure 1-1 is not static, it evolves dynamically, and that allows for the rise, development, and damping of local celestial effects. That is, the birth and death of stars seems to go along with changing local values of the Space and Time curvatures. In one example, when a star collapses under its own gravity, the increase in gravitational attraction due to the increase in proximity of its matter contents, corresponds to the "fastness of passing time" in that region to become slower. This translates to point S in figure 1-1 shifting to the right in the direction a' → b. Based on material to follow, once such shifting of point S gets to reach and cross the nodal point b, this leads to conventional matter decay into particles of negative energy, in which case a gravity-powered repulsion leads to a supernova explosion.

In the next sections we shall elaborate on how the curvature of the time and space metrics -or of the negative reciprocals of those- gets to associate to different fields, like the gravitational, the electric, or the intrinsic-magnetic. It shall arise that fields with radial-like "field lines" have their field lines follow the loop of the axis XX' of figure 1-1, of which we may directly perceive only the real portion, while the remaining portion exhibits properties that may be described by imaginary terms. Likewise, we shall consider how opposite-energy versions of the conventional fields engage in effects of reaction (in the form of imaginary coordinates of these fields).

4

Cosmic Regions & their Particle Inhabitants

The phase-shift between the Time and Space curvatures shown in figure 1-1 leads to the creation of 4 distinct regions, as follows: Region d' - b which corresponds to the classical universe where both the space and the time metrics have positive values, region   b - b' where only the space metric is negative, d - d' where only the time metric is negative, and b' - d where both the space and time metrics are negative. Each of those regions constitutes natural habitat for different kinds of particles. For reasons of simplicity we shall here refer only to leptons (e.g. the electron, positron, neutrino, antineutrino), leaving composite particles for later, since they are composed of quarks which constitute more perplex cases and we shall address those in section 10.

i. Region d' – b

This region coincides to the known universe where galaxies constitute stable habitat for elementary particles made of "**positive matter**", like the electron. The Schrödinger equation which describes electrons, relates **a second space derivative ($\partial^2\Psi/\partial x^2$) to a first time derivative ($\partial\Psi/\partial t$)** for calculation of a particle's matter wave. According to conventional theory, the existence of mass bends the surrounding spacetime, and this bending corresponds to a particle's gravitational field. As we shall see ahead, however, the notion of this bending being a "consequence to the environment of the particle's existence", is not that accurate. It is more proper to consider that each elementary particle's own gravitational field (which contributes to the gravitational field of a larger object) constitutes part of that particle's very essence, and that appears to reflect in the imaginary portion of the particle's wave function. Under a simplistic pattern of thought (to be further considered later), the particle's wave nature arises by relating the wave function's **"second space" derivative** to a **"first time" derivative**, and that, in turn, is counterbalanced by an inverse analogy with respect to the particle's surroundings: That is, the gravitational field is accounted for in meters/sec^2 which we shall loosely represent as dx/dt^2, which relates a **"first power" of space** (dx) to a **"second power" of time** (dt^2). Actually, that the units of the gravitational field are same as the units of conventional acceleration, is not exactly accurate either. While these particular units of the gravitational field work well in calculations, this is happening as a consequence of the principle of equivalence. Instead of describing a change in velocity with respect to time (dt), the units of the gravitational field should rather account for a curvature, like a "difference in the fastness of passing time" Δt, with respect to a change in altitude dx. In fact, as we shall see in section 5.1, the curvature in the metric of *time* allows for an intuitive explanation on the functioning that gets the gravitational field to accelerate matter.

Furthermore, while according to existing theory space and time are considered tightly interrelated entities, it appears that this is not too accurate either. Since space and time may be subject to a phase shift between them as introduced above, this suggests that they constitute distinct entities, much like voltage and current are distict entities in electronics, despite their tight interrelation too. In such case, the gravitational field should rather associate closely to either one of them, in this case to the curvature in the metric of **time**, as experimentally confirmed. As we shall later consider, the curvature in the **space** metric has a complementary role, which applies in a symmetric way.

As we move on we shall actually figure that the curvature in any metric in general corresponds to a potential. Just like the gravitational field (which has radial-like field "lines") corresponds to a curvature in a metric, it should be reasonable to expect that a similar condition should hold for the electric field as well (which also has radial-like field "lines"). In fact, based on material to follow in section 5, the electric field will be shown to associate to curvature in a variant of the metric of space, which involves a particular reciprocity. And in relation to that, in section 8 we shall consider how properties associating to a particle's charge may be approximated through a "complementary-to-the-Schrödinger" equation, which relates a second *time* derivative to a first negative *space* derivative (the inverse to what holds for matter).

ii. Region d - d'

In this region of figure 1-1 the TIME metric takes negative values, where, as explained earlier, negative time should be perceived as a form of negative energy, with events in that region following causality given the negative energy conditions, where that has nothing to do with the notion of "events evolving backwards" like in a movie run backward. The Schrödinger equation that should derive the particles of that region should rather relate a **second space derivative ($\partial^2\Psi/\partial x^2$) to a first negative-time derivative ($\partial\Psi/\partial(-t)$).** The matter-essence of particles that inhabit that region should gravitationally force and accelerate other matter as per $(dx/d(-t)^2)$, instead of the "conventional" gravitational acceleration (dx/dt^2). But since both these terms bring an equivalent result $(dx/d(-t)^2)=(dx/dt^2)$, the said particles should respond the "usual" way to an external gravitational field, so they should be gravitationally attractive (just like conventional matter is). According to material to follow, it similarly applies for the inertial properties of particles of that region, since a change in their momentum is resisted in the same way to as the change of momentum of conventional matter is. It arises that the type of matter referred to in this case corresponds to **antimatter**, ascribing to particles like the positron, which is predicted to be gravitationally attractive,[6] as when an antimatter particle is produced it falls "down". And, considering it is gravitationally attractive, we shall assume that antimatter constitutes "matter of positive sign" (just like regular matter does).

Along these lines, the region d - d ' constitutes favorable ground for antimatter particles, with such particles reaching stable form within that region. As for the charge of these particles, we shall later see that this appears possible to account through the aforementioned "complementary-to-the-Schrödinger" equation, where the involvement of an opposite sign on the time metric reflects on the particle having an opposite sign of charge, than that of the electron.

Notice that while antimatter reaches formation in the region d − d ', it should also reach formation in our classical region d ' − b, and respond to gravity in meters/(-sec)^2, or loosely $(dx/d(-t)^2)$, where the power of two makes its gravitational behavior same as that of conventional positive matter. However, since a positron lives in negative energy in time, when this particle is produced in our cosmic region where positive time prevails, its opposite energy in time makes it vulnerable to match with a similar amount of energy in positive time and annihilate, so free positrons interact with electrons and annihilate together.

The above notion may allow to address the as-of-now unresolved issue of cosmology concerning **the abundance of matter over antimatter** in the known universe, which holds despite the fact that in accelerator experiments it is observed that antimatter is produced in pairs in equal quantities with matter. So, one would normally expect to have equal abundance of matter and antimatter in our conventional universe. That however is not the case, as antimatter is not stable in our classical region as mentioned above. While instead, it should be stable in regions of negative time metric, where conventional matter should be unstable and susceptible to annihilation. In relation to that, a region like d − d ' where the time metric is negative should contain an abundance of antimatter over matter. These antiparticles should actually engage in forming atomic and molecular formations of antimatter within that cosmic region, which should in turn gather in larger, celestial formations (same as it applies for conventional matter in our cosmic region). Since the space metric in the region d – d′ is positive, such kind of celestial bodies (composed of antimatter) should be visible from earth. In fact, indication or evidence of celestial objects constituting of antimatter exists, on the basis of the prediction that light from there sources should have negative index of refraction, which is observed via telescopes with concave lenses.[7],[8]

Later, in relation to figure 4-2, we shall additionally address why it appears to be lesser amount of antimatter (and a corresponding lesser number of such exotic celestial structures made of antiparticles) in nature, than conventional celestial objects. This shall be possible to do through comparison to an effect of energy transfer in electronic circuits. As we shall see, the reason associates to the lesser total energy contained within the relatively short cosmic region d – d′ than the energy content within the relatively wider region like d′ – b.

iii. Region b - b′

In this region, the SPACE metric takes negative values (with respect to our classical world point of reference). A Schrödinger equation describing particles of that region should relate a **second negative-space derivative $(\partial^2\Psi/\partial(-x)^2)$ to a first time derivative $(\partial\Psi/\partial t)$**. Matter inhabiting that region should associate to curvature of the local spacetime metrics that causes gravitational-like acceleration of the form $(d(-x)/dt^2)$. Since $(d(-x)/dt^2)=(-dx/dt^2)$, this differs in sign from gravitational acceleration, therefore the corresponding wave function should describe a particle of negative matter, being REPULSIVE to conventional matter (hence, different to antimatter which is known to be gravitationally attractive). That is, it describes a type of *negative-matter monopole*, which we shall here postulate to concern the negative energy solutions of the Dirac relativistic equation. As that particle has negative energy, it should be subject to a quantitative inversion, having the cosine coordinate of its wave function being imaginary, and the sine coordinate being real, the opposite of what holds for conventional matter. This inversion ascribes to a particle that should rather move and interact superluminally within its corresponding cosmic region, while in our classical universe it should exhibit imaginary nature. That imaginary nature does not allow it to take ordinary form and interact gravitationally with conventional matter, neither it exhibits inertia, thus it appears as being massless. This type of particle appears to correspond to the **antineutrino** (rather than the positron [9]).

Antineutrinos (and neutrinos alike) should be considered negative-energy particles, or rather, "negative-matter" monopoles. Within their own cosmic habitat of region b - b′ they should interact among each other gravitationally through a corresponding "**negative gravitational**" field. This should be attractive between antineutrinos (and neutrinos alike), since they all possess same sign of matter, of negative sign. Likewise, within their cosmic habitat antineutrinos should exhibit conventional inertial properties (just like electrons do in our own cosmic habitat). However, their negative-matter essence (which requires imaginary terms to account for) prohibits them from directly interacting gravitationally with conventional matter. Therefore, in the cosmic region where we live in, the antineutrinos appear to be massless.

Rather symmetric conditions seem to hold for these particles' charge. While antineutrinos' essence should be compatible to superluminal conditions (accounted for by imaginary terms), they still need to carry a form of charge for reasons of energy balance which shall be discussed in section 9. The charge carried by antineutrinos concerns an imaginary (negative energy) version of the conventional electric charge, which we may refer to as "**anti-charge**". This shall appear to relate to the "weak charge" mediating the weak interaction. The anticharge comes along with its corresponding "**anti-electric**" field, which applies as a conventional action field within the cosmic region b - b′, while with respect to the classical world this charge and field are imaginary (accounted for by imaginary terms) which is why the antineutrinos (and neutrinos alike) illusively appear to us as being neutral particles.

Within a cosmic region b - b′ antineutrinos should interact much like electrons do in our own cosmic region, and should engage in atomic formations around nuclei composed of negative baryons. Corresponding molecular formations, in turn, should be expected to settle in celestial-scale objects. To view such objects does not necessarily require to dive to a black hole past the surface corresponding to point b of figure 1-1. Such objects should be possible to trace in transitioning celestial effects, where the prevailing values of the metrics of space and time shift due of local evolution of the corresponding transition. One such case is when very big stars burn out their fuel, and gravity gets them to collapse. During this process gravitational action intensifies, as if the value of the metric of time shifts in the direction a′→b′ where the slope of the time curvature progressively gets steeper. When the shrinkage reaches a critical limit associating to the Pauli exclusion principle (the Chandrasekhar limit), then its value appears to cross over the nodal point b (toward the right side) where the space metric takes negative values. The crossing of the nodal limit gets conventional matter turn to negative matter (neutrinos) through inverse beta-decay, which is gravitationally repulsive to conventional matter. Through this transitioning process, positive and negative matter get to co-exist in particle form, in very close proximity (allowing them to interact in close range), in which case the gravitational repulsion between these two forms of matter constitute the key reason that a **supernova** explosion takes place. The remnant should concern a new type of celestial body where electrons have turned to neutrinos and protons have turned to neutrons. What we are left of, appears to be a celestial object constituting a local cosmic region b - b′, whose particles should exhibit certain superluminal properties, which may comply to observation.[10],[11] A similar process appears to takes place when (negative) matter is gravitationally attracted in the direction c′→d′ and space curvature crosses over the nodal point d, where that should yield an inverse process that emits light and appears to refer to what holds for pulsars which should emit Z^0 bosons within their cosmic region, behaving like photons do in our own cosmic region. Also in that region, the electromagnetic interaction should produce photons interacting just like Z^0 bosons do in our own neighborhood. These photons should play dominant role for the high luminosity of these celestial objects.

On a note of significance that will be particularly addressed in the next two sections, the "anti-electric" field that antineutrinos should carry (acting as a normal field within the region b - b′, while with respect to our world this field has imaginary essence), will be shown to relate to a reaction version of the electric field that is involved in driving the electromotive potential in our cosmic region (there engaging in the form of displacement-electric field D). Likewise, the analogous "negative gravitational" field that antineutrinos carry (acting as a normal field within the region b - b′, while with respect to our world it has imaginary essence), will be explained to relate to a reaction version of the gravitational field that is also involved in generating effects of vortex flow of fluids in our own cosmic region (there engaging in the form of a displacement-field, which shall be considered in section 6).

iv. Region b′ – d

In that cosmic region BOTH the space and time metrics are negative. A Schrödinger equation that describes the particle inhabiting this place should relate a **second negative-space derivative ($\partial^2\Psi/\partial(-x)^2$) to a first negative-time derivative ($\partial\Psi/\partial(-t)$)**. The particle derived in this case seems to correspond to a negative-matter particle of opposite charge than the antineutrino's, and that refers to a **neutrino**. As it possesses matter of negative sign, it should be gravitationaly repulsive to conventional matter. However, its negative-matter essence should have its wave function accounted for by imaginary terms, and that prohibits it from directly interacting gravitationally with conventional matter, as well as from exhibiting properties like conventional inertia. Therefore, in the cosmic region where we live, the neutrinos appear to be massless, while instead, within their natural habitat of region b′ – d they should gravitationally attract each other, and should also attract antineutrinos for the same reason that electrons gravitationally attract positrons. In their cosmic habitat of negative space, neutrinos should be moving superluminally, while for interaction with conventional matter they would need to slightly cross the border from negative-to-positive energy in order to interact with positive curvature, and this is the reason why only left-handed neutrinos are found to interact in our cosmic region, due to their marginally lower energy which allows them to interact in the classical side of the universe.

By symmetry to what was described for antineutrinos, neutrinos carry an opposite sign of anti-charge which corresponds to "**negative anti-charge**". This again, is not appearing towards our classical world due to its imaginary nature, so the corresponding particles appear being neutral. However, in their own habitat they should interact based on that type of anti-charge which they carry. According to material to follow, anti-charge refers to the "charge" involved in the so-called "neutral currents" of the electroweak interaction. Its associated anti-electric field also has imaginary nature, so it should damp out and become evanescent in extremely short range within positive space (the classical world). This is also a reason why the weak interaction has a very short range of action in our classical cosmic region.

Furthermore, for same reason that the positrons are able to reach particle form and interact in our classical universe, but with short half-life, neutrinos should be able to reach particle form and interact in the cosmic region b - b′ but with short half-life. And likewise, antineutrinos should be able to reach particle form and interact in the region b′ − d but with short half-life.

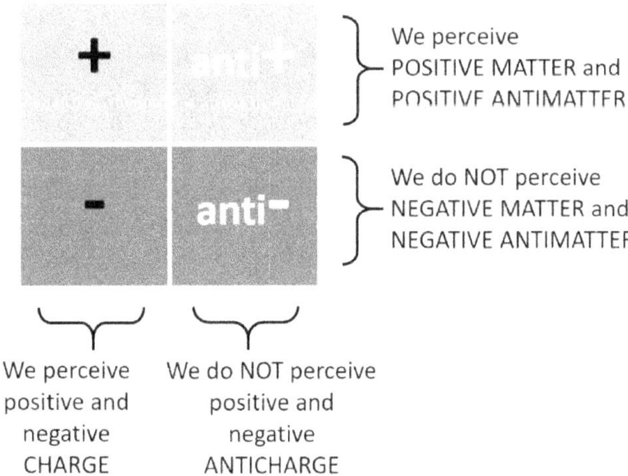

Figure 4-1. Orthogonality between matter and charge, indicating which rows & which columns we directly perceive, and which ones we don't

On the basis of the above notions, the particle inhabitants of all four spacetime regions depicted in figure 1-1 (regions b-b′, b′-d, d′-b, d-d′) occupy the same cosmos that exists all around us. In our own cosmic region, given the positive sign of the metrics of space and time, only certain particles may reach stable form, as indicated in figure 4-1. Speaking of leptons, we may directly perceive **positive matter** (electrons) and **positive antimatter** (positrons) and we may not directly perceive negative matter (neutrinos) and negative antimatter (antineutrinos). Likewise, in terms of charge, we may directly perceive **negative and positive charge** (of electrons and positrons), and may not directly perceive negative and positive anti-charge (of neutrinos and antineutrinos).

On a point of significance, even though we may not directly perceive the fields of certain of these particles (neutrinos and the antineutrinos), in the next sections we shall figure how their particular fields (which have imaginary-field nature) are involved in our everyday life in **reaction potentials**. Like for instance, in affecting inductive currents via the electromotive potential. This is possible via interaction between a displacement-field version of these fields, with conventional particles' imaginary part (to be specified later). In such case, the interaction of two imaginary entities may yield a real negative outcome as per the relation $i^2= -1$, where the negative sign signifies reaction.

Notice that the sum of all the energy associated to negative matter and anticharge which lie within other cosmic regions (constituting the particles that inhabit these regions), plus the energy in these particles' fields, as well as the energy of such fields when involved in affecting reaction potentials on conventional particles in our cosmic region, may be considered to constitute a significant part of the so-called **dark energy** in the universe. However, the notion of dark energy and the amount of it in the cosmos needs to be re-considered with caution. The reason is, that this type of energy was originally conceived (actually hypothesized) in order to match and explain the accelerated expansion of the universe. While, under the present model the accelerated expansion of the universe is attributed to a different cause, which concerns the opposite slope of fields like the gravitational in the range of the very distant universe, as indicated through the curvature of the metrics in figure 1-1. Furthermore, negative matter and its corresponding fields deploying over the non-classical cosmic regions engage in interactions which rather lie beyond what dark energy was conceived to address for in the first place.

<u>**A parallel to electronics:**</u> **(Abundance of matter over antimatter)**
It is of interest to make a parallelization between the phase shifts of the Space and Time waviness shown in figure 1-1, and the Voltage and Current

oscillations in electric circuits, especially with regard to the flow of energy in each case. This is done through figure 4-2, where the upper part illustrates an oscillation of Voltage (black curve), an associated oscillation of Current (dashed grey curve), and the power curve (doted curve) which concerns the product of voltage times the current. During the time that the oscillations of voltage and current have both positive or both negative values the power is positive, which means that **energy is transferred from the capacitor to the inductor**. While, during the time that either one of the voltage or current is positive and the other is negative, the power is negative which means that **energy is transferred from the inductor to the capacitor**.

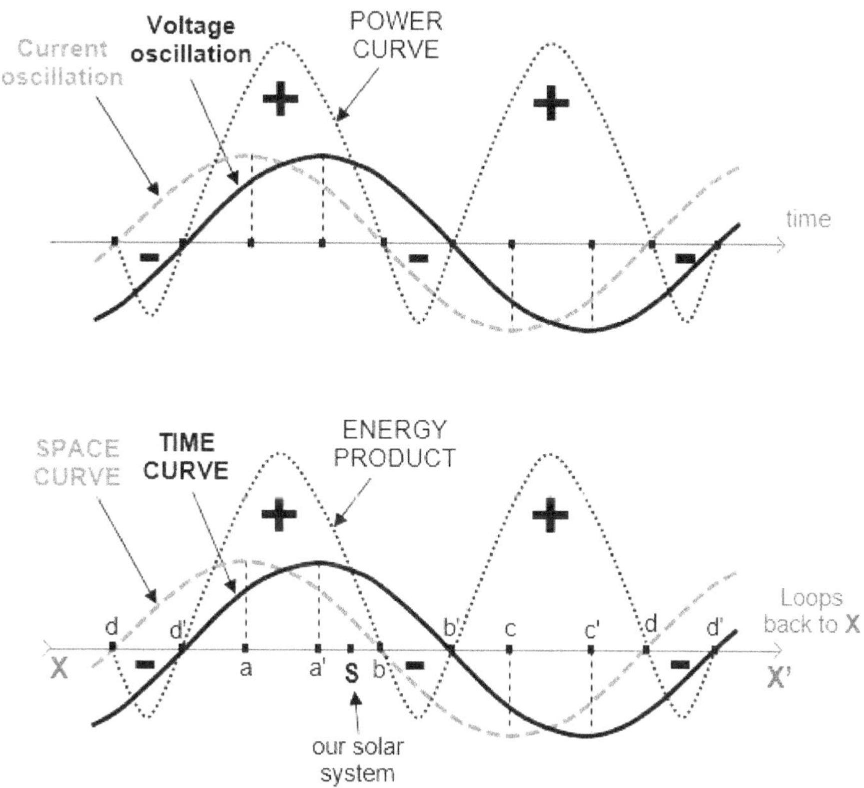

FIGURE 4-2: Waviness of power, exhibiting large positive antinodes and small negative antinodes, in an electrical case (upper), and in a mechanical case (lower)

A similar type of analogy holds in the mechanical case (see lower part of figure 4-2) which concerns the energy contained in the curvature of the Space and Time metrics. Here the phase shift between the wavinesses of the space and the time metrics deploys over space (instead of developing over time which was the case in electronics). Therefore, the flow of energy is not changing depending on the progress of time, it is instead changing with respect to the spatial positioning within the cosmic scale. The particular shape of the "energy product" (dotted curve) provides information as follows:

-   **Region d′ – b:** This region corresponds to our own habitat, and accumulates a large amount of positive energy (re. the area under the dotted curve which have a positive sign) compared to its neighboring negative energy regions (areas of the dotted curve with a negative sign). This translates to large amounts of conventional matter accumulated in this region of positive energy, an associated abundance of celestial bodies made of particles of conventional matter, and a large amount of energy in action fields (e.g. the gravitational) engaging conventional matter.

    And just like in electronics the positive power concerns transfer of energy from the capacitor to the inductor, in the context of mechanics we have flow of energy from the positive energy side toward the negative energy side. This takes place through acceleration caused by action fields (e.g. the gravitational), where the energy involved in accelerating matter exceeds energy involved in affecting impedance (like inertial impedance where energy is consumed without doing work). A similar energy flow may concern the displacement of matter in vortex-like effects that shall be addressed later.

-   **Region d – d′:** The short extent of this region (in comparison to the larger extent of the classical region d′ – b of our own habitat) translates to the area of negative antinode of the dotted curve (marked with "–") being lesser than the area of the positive antinode (marked with "+"). This indicates that the particular cosmic region of negative sign accumulates a lesser amount of energy in the form of **antiparticles** in that region of the universe along with their associated fields (e.g. the gravitational). These antiparticles may conglomerate into celestial formations of antimatter. The substantially lesser amount of energy contents in that cosmic region (with respect to the region of our own habitat), translates to a considerably lesser amount of antimatter particles, and a corresponding lesser number of celestial bodies in existence comprised from antimatter (with respect to celestial bodies made of ordinary matter, which come in abundance).

Note that celestial bodies in a cosmic region like d – d' should be traced in astrophysical observations indicating gravitational attraction in that opposite direction, due to the opposite inclination of the external gravitational field). Light coming from these sources is predicted to have opposite index of refraction, and indication or evidence of celestial objects of such property of light (therefore appearing to constitute of antimatter) is observed via telescopes with concave lenses.[7],[8]

Regions of negative time metric in the universe could be from extremely large, to extremely small. An example of an extremely large region would concern one that houses celestial bodies made of antimatter. In the other hand, an example of an extremely small region could concern even a single antimatter particle that came into existence within our own cosmic region (of positive time and space metrics). Even though such antiparticle (for example, a positron) may come into existence through a particle-antiparticle pair production, it still constitutes (by itself) a small spot of negative energy in time. It's like constituting a minimal d – d' region, a tiny "island" of negative time within our cosmic region. When such a region is as small as a single particle, however, then the surrounding area of positive time does not constitute favorable place for such an antimatter particle to maintain in existence. This is so, since the negative energy in time concerning the antiparticle, finds ample availability of particles of corresponding energy state in positive energy in time (in the form of ordinary matter particles like electrons) all around it, and that makes the antiparticle vulnerable to combine with conventional matter of equal energy in positive time, so as to annihilate together.

While the cosmic region d – d' should be inhabited by antimatter, if ordinary matter becomes present in that region it should be subject to gravitational repulsion by celestial bodies whose outer surface corresponds to point d of figure 1-1 (as viewed from its right side). The corresponding repulsion on these places is due to the metric of time having negative values in that region (and not due to antimatter being gravitationally repulsive, which is not the case). This process may leak negative-energy-in-time toward the positive energy side.

Furthermore, another process that might leak energy toward the opposite energy side, may concern negative matter (instead of antimatter) involved in transitional effects of reaction where the negative energy in reaction overrides the energy in action forces. For example, in vortex-like flow, which will be explained to concern a mechanical equivalent to the electric counterpart of an induction current in a metal ring. There, we may have an opposite flow of energy through the interaction that involves the imaginary portion of particle's wave nature (to be addressed later).

- **Region b – b′**: This region exists outside a black hole's event horizon, and seems to associate to the predicted existence of an ergosphere around a rotating black hole. According to the new model considered here, however, conventional leptons like electrons and positrons should not be able to reach stable form in this region due to the space inversion that takes place at point b. While instead, this region should constitute stable habitat for **negative-matter particles**, like antineutrinos (if we speak of leptons), where they should behave similarly to as conventional matter (e.g. electrons) do in our own cosmic region. They should exhibit inertia, and their atomic/molecular formations should become involved in vortex-like processes, leaking energy toward the opposite energy side under a similar way to as conventional matter does through effects of reaction in our own cosmic region. The negative-matter particles of this region should also conglomerate into celestial formations. However, the relatively short extent of the corresponding region, and more particularly the short *area* of its negative energy antinode (dotted curve), transalates to a lesser amount of negative-matter particles in existence. Hence, a correspondingly lesser number of celestial formations made of such negative form of matter should be in existence (within such cosmic regions) in the universe, with respect to celestial bodies of ordinary matter which come in abundance.

Regions of negative energy in space could be sized from extremely large, to extremely small. An example of extremely large regions may concern areas outside rotating black holes, housing celestial bodies made of negative antimatter (like antineutrinos, if we speak of leptons). In the other hand, an example of an extremely small region could concern even a single negative-matter particle. If such a particle could be able to reach form in our own cosmic region, it would constitute a small spot of negative energy in space, like a minimal b – b′ region, a tiny "island" of negative space within our cosmic region of the positive space metric. According to material to be presented ahead, however, negative-matter particles cannot reach form within our cosmic region of positive space, and this is because their negative matter essence has imaginary character with respect to the point of reference of our classical world. There is however an exemption that allows for the existence of negative matter in our cosmic region. In relation to material that will be presented in section 10, quarks (the constituents of nucleons) have a particularly unusual essence: As it shall be addressed later, the Weinberg angle in that case gets part of their nature belong in negative energy, and another part in positive energy. In a way, a quark gets to correspond to 1/3 or 2/3 of an electron or a positron, and 2/3 or 1/3 of a neutrino or an antineutrino. Quarks may collectively meet certain criteria of quantization, which allows them to reach stable form in our part of the universe, so the negative matter portion of each quark may stay stable and behave as a tiny "island" of negative matter. And as we shall figure later, the existence of this negative matter portion of quarks within our universe, along with the fact that quarks are in large abundance throughout any galaxy (being located within the nuclei of all atoms), has their negative energy portion participate in a negative energy interaction, which causes leakage of energy from the negative energy side toward the positive energy side. That particular interaction shall be described to cause a potential that seems to resolve the rotational behavior of galaxies, without the need of existence of the so-called dark matter. Or, in an alternative way to look on the same point, the negative energy portion of quarks may be considered to concern the missing dark matter which becomes involved in a negative energy interaction, as shall be addressed in section 5.4, point 6.

- **Region b′ – d**: This is a significant region, because the "energy product" of the *negative* energy in time-curvature multiplied by the *negative* energy in space-curvature yield a *positive* outcome. That signifies that interactions in that region may deliver actual work in our classical region (instead of affecting impedance). This also suggests that energy may leak from the imaginary domain toward the real domain, with interactions of reaction character yielding an enhancing outcome (instead of an impeding one). In such cosmic region the main inhabitant leptons appear to be the neutrinos, while -in relation to material to follow- celestial objects that form in such cosmic region may concern **quasars** and **blazars**.

It is of particular interest that interactions which yield positive outcome toward the real world (due to the co-involvement of both negative energy in time as well as in space) may be traced at the microscopic/quantum scale, via the engagement of the imaginary portion of particles' wave nature. This is possible due to the quasi-particle properties of quarks that were mentioned above and shall be analyzed ahead, where a portion of every quark interacts in positive energy and another portion interacts in negative energy, concurrently. The engagement of the negative energy portion may result in producing an enhancing (instead of impeding) outcome. To explain the process, it requires to present aspects that will be introduced in section 10. Such enhancing kind of behavior at the microscopic scale may be found in both electronics as well as in mechanics, allowing reaction fields to serve as a source of lasting energy.

In electronics it will be shown to be responsible for the lasting potential across the poles of an **electric battery** (unlike what holds with charge between the plates of a capacitor, where the energy gets quickly consumed). While in terms of mechanics, a potential energy of this type will be described to be the one that powers **exergonic bio-chemical reactions** which provide lasting energy, that powers life. A corresponding process might also be involved in providing the energy surplus needed for the conversion of protons into neutrons inside stars.

An interesting question is whether it's the cosmic waviness that dictates the housing of particles per region, or is it the particles that dictate the waviness. Based on what is to follow in the coming sections, the answer to this "chicken-n-egg" problem is rather straight. Since either of these two processes affects the other, they could only go together, concurrently, and in mutual balance. After all, this is what allows for the total positive and the total negative energy, of both real and imaginary coordinates, in the cosmos to sum up to null, making it plausible for the cosmos to exist without any need for external supply of energy, and without the need for the existence of a "cosmological constant".

5

Basic Concepts about Fields and How They Apply

5.1 The Gravitational Field (g)

We shall start with an insight on the essence of the gravitational field, as well as on the process through which it gets to accelerate matter. According to the theory of relativity, the existence of mass curves Spacetime. The curvature of the metric of Time and the curvature of the metric of Space are tightly correlated to each other, however we shall here focus on the metric of Time. As experimentally confirmed, the existence of earth's mass curves the metric of time, and because of this, a clock at sea level tics slightly slower than a clock positioned at a higher altitude (see figure 5-1, left part). In this case, we shall presume that the difference in the "fastness of passing time" between two adjacent points (the time metric having slightly different value at one spot than it does in an adjacent spot at different elevation) constitutes a **gravitational potential**. Notice that while the curvature of space has a complementary role, we shall later figure that space has a differentiated involvement.

Based on such presumption, let us see how a gravitational field may cause acceleration of matter. To figure that out, we shall use a pattern of though as follows: As per a textbook paradigm of the theory of relativity, when a rocket (or any material object) accelerates, time progresses slightly faster in its front part than it does at the rear part, as illustrated in figure 5-1, right part. This is due to an effect of relativity which resembles the Doppler effect taking place in the time domain, according to which, a hypothetical clock positioned in the front part of the accelerating rocket moves slightly faster than another clock positioned in the rear part of it.[12] Assuming not-relativistic velocities where v^2/c^2 tends to zero, the difference in time fastness has a factor of $(1+gd/c^2)$

where g is the acceleration and d is the distance between the two clocks. This difference is obviously tremendously small, yet it does exist between the front and the rear parts of any accelerated object.

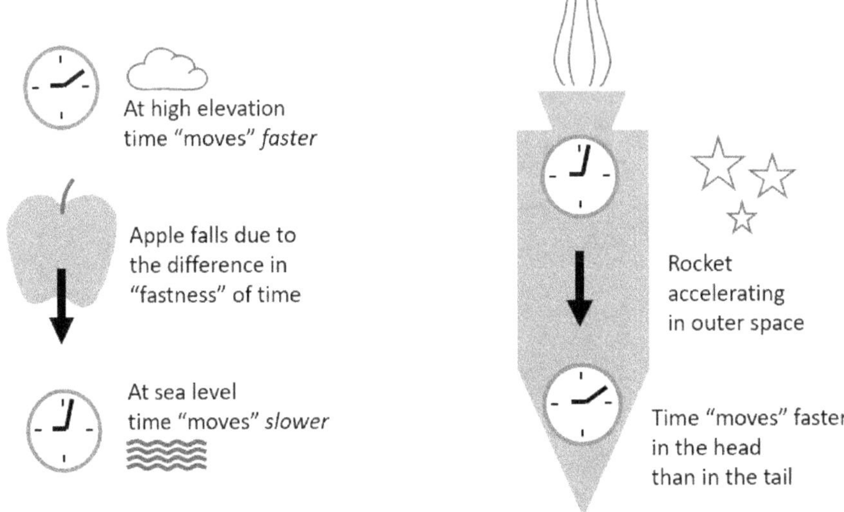

Figure 5-1: Explanation of how the gravitational force applies: Time "passes" at different rates, here is what holds for gravity (left) and for acceleration (right). These two processes counterbalance each other, so an object accelerates in order to relax and release the gravitational potential.

It turns out that the difference in the fastness of passing time due to acceleration (e.g. while falling), points in the opposite direction, and balances out the difference that the object experiences in the fastness of passing time due to the (external) gravitational curvature. Actually, due to the principle of equivalence, the same amount of difference in time ($1+gd/c^2$) applies between two clocks in a gravitational field g, separated from each other by the height d.

In relation to the above, the difference in the "fastness of passing time" attributed to an external gravitational field, should be treated as a potential difference, having a corresponding potential energy associated with it (just like a compressed spring also possesses potential energy). And when an object is let to fall toward the ground, the acceleration-related curvature in the time metric (time moving FASTER at the part of the object heading toward the ground due to the Doppler-like effect in time) exactly counterbalances the gravity-related curvature in the time metric (the time in the environment passing SLOWER toward the ground due to the earth's gravitational field). Consequently, for as long as the object KEEPS accelerating (while falling) the difference in the fastness of time associated to the acceleration exactly balances out the difference in the fastness of time due to gravity. And as a result, the potential energy associated to the gravitational curvature in time gets released/relaxed. And indeed, while an object free-falls, it feels weightless, just like in the example of astronauts' training inside an airplane that follows a parabolic flight path. From the point of view of the falling object, its acceleration releases the gravitational potential exerted on it. This release of potential in the time domain is what drives the acceleration to develop by its own, for similar reason to as a compressed spring would stretch out all by itself in order to release its compression.

Following the above, and in relation to what we shall further consider in section 7, the statement that the gravitational field corresponds to the curvature of "spacetime" attributed to the presence of mass, is not accurate. In fact, **that the time curvature and the space curvature can be subject to a phase-shift between them, implies that these two behave as distinct entities** (much like oscillations of "voltage" are distinct to oscillations of "current" in electronics, despite their tight correlation). Consequently, the gravitational field should associate to either one of these curvatures, not both. In such case, it arises that it associates to the curvature in the metric of time, as experimentally confirmed. Furthermore, we shall later figure that each particle's own gravitational field (contributing to a total gravitational field in a region) should be understood to comprise an integral part of the particle's own essence (and not a consequence to the environment due to the presence of this particle's mass).

Regarding the field lines of a particle's gravitational field, these look radial-like and open-ended, and their particular trail designates the path toward which the "fastness of passing time" changes (since time moves slightly faster the further away from the particle or larger material object, as an effect of gravity). This feat, however, does not go without a limit. While the object's gravitational field lines appear to head toward infinity, this does not mean that the fastness of "time passing" keeps ever increasing, in fact it neither becomes flat after some point very far away. Instead, at some range at mid-distant universe time stops moving faster, it reaches a peak, and then starts moving slower and slower, even though the gravitational field lines still point toward infinity. In such way, the fastness of time attributed to each particle's gravitational field follows a *waviness*, as illustrated schematically for the Time curvature in figure 1-1. This also means that at some point really far away (a hypothetical "event horizon" referring to point d'), the fastness of time reaches zero, and beyond that range the time metric appears to turn negative. Note that, even though the gravitational field of a fermion looks radial-like, it shall later arise that each field "line" actually forms a hypothetical loop, of which we directly perceive the real portion (which is radial-like), and we do not perceive an imaginary portion. Yet, we shall later figure that the imaginary portion does engage in effects of reaction, taking place all around us. Furthermore, in section 7 we shall additionally consider how the gravitational field relates to the imaginary part of a particle's wave function.

In terms of units, gravity shares the same units with acceleration, both being measured in meters/sec^2 (referring to dx/dt^2). This however is not accurate either. The units of the conventional acceleration describe a change in velocity, and this works for the units of the gravitational field due to Einstein's principle of equivalence as the gravitational force resembles acceleration of matter. However, these particular units do not reveal the relation of the gravitational field to the difference in the values metric of time (meaning that time passes at different fastness) between adjacent points in space, which constitutes the actual cause of gravity. That difference in the metric of time could be accounted for indirectly through the current units only if we loosely interpreted these units on the grounds that the differential in the metric of time Δt between adjacent points over space is counterbalanced through a change in velocity dx/dt until reaching this adjacent point (where the time metric has a slightly different value). Other than this approach, in the next section we shall also figure that acceleration could also be represented as a "change in the flux of matter" (not mass). And the reason that this association is said to concern "matter" (instead of "mass"), is that mass rather refers to the impeding behavior with which matter is "dressed" with, due to the existence of the phase shift shown in figure 1-1. Thus, mass is a relative quantity, depending on the value of that phase shift, while matter (which concerns an accumulation of quantized units of energy) concerns a rather specific measure of amount of energy.

While above we have referred to acceleration arising from the gravitational action, mechanical acceleration/deceleration may also be driven by other mechanical forces, like for example forces associating to pressure (e.g. the Bernoulli pressure) or forces associating to particles' mechanical intrinsic spin properties, which shall be considered in the next sections. There, we shall figure how motional properties of matter engage in the generation of potentials (like the one driving the Bernoulli pressure), and how such forces differ to the gravitational. As for the gravitational behavior of **antiparticles** (like the positron), the gravitational force in that case applies the same way: The essence of antiparticles incorporates an opposite sign of progression of time as stated earlier, however since the gravitational action involves the power of two, where, meters/(-sec)^2 = meters/sec^2, antiparticles respond the "same" way to an external gravitational field, therefore they are gravitationally attractive to ordinary matter. In that sense, the gravitational field of antiparticles corresponds to ordinary "positive gravitational field".

In relation to material to follow, a different, negative energy version of the gravitational field also exists, which ascribes to "negative sign of matter", which is completely different to "antimatter" and associates to an imaginary version of matter which refers to the essence of neutrinos and antineutrinos. The corresponding negative energy version of the gravitational field should act as an ordinary "real" field within cosmic regions where the space metric has negative values. Such regions should constitute natural habitat for neutrinos and antineutrinos, where these particles should behave the same way as ordinary matter does in our own cosmic region, and should gravitationally attract each other within that habitat. However, as the negative gravitational field is imaginary with respect to the classical world, neutrinos do not interact gravitationally with conventional matter (except in marginal conditions which will be considered in section 9), so they appear to us as being massless.

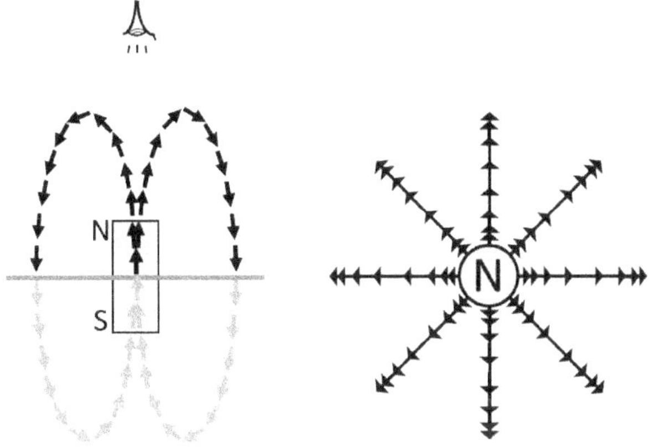

Figure 5-2: The gravitational field looks similar to the field of a bar-magnet when viewed from above the one pole, and that makes it look radial-like.

Let us now look further into the gravitational field "lines" of an electron. Their radial-like shape should incorporate the waviness in the fastness of time illustrated in figure 1-1. That is, each field "line" follows the trail of change in the fastness of time passing, which moves faster and faster until mid-distant universe, and from that range and on the time metric starts having diminishing values, reaching zero value (time freezes) at the nodal point d′, then turning negative and eventually becoming imaginary and hidden from perception, until the field lines complete a hypothetical loop that reaches back to the particle. Thus, the portion that we perceive from each gravitational field "line" is *roughly* the positive antinode of the total waviness of the complete field line's loop. (A special condition also applies by the nodes, but shall be addressed later). In relation to that, the gravitational field lines of a point particle may be imagined to look similar to the magnetic field lines of a bar magnet (figure 5-2 left part) **when viewed from above the one pole** (figure 5-2 right part). This "top-side" view illustrates that the magnet's field lines may look radial-like and open-ended even though they constitute loops. It's like projecting their 3-dimensional trail to a 2-dimensional plane (therefore "missing" the 3rd dimension, of depth). That "missed" dimension, however, could be possible to represent by changing the density of field "arrows", as shown at the right part of figure 5-2. This change in density of arrows may represent the incorporated change in fastness of time (which the plain field lines do not portray).

As per the above, we may loosely imagine an electron as a sort of unconventional gravitational-dipole (thus, an unconventional matter-dipole as well). We may perceive only the real pole of that dipole, and the corresponding real part of the gravitational field (black arrows in the upper portion of the left part of figure 5-2), and we may not directly perceive the imaginary portion of the gravitational field (the grey arrows at the lower portion of the same figure) which re-attach to the particle through an "imaginary **negative pole**".

A significant difference between the radial-like gravitational field, and the radial-looking "view from above" of the magnetic field, has as follows:
- In the gravitational case, the field "lines" **spread** radially away, all the way until the very distant outer space (point d′ of figure 1-1, where they turn negative, and later turn imaginary at point d).
- In the case of a magnet's pole viewed from above, the field "lines" do not spread far out as they quickly (in short range) converge back toward the other pole of the magnet.

This is so because in the magnetic case the "other pole" (which is hidden behind the pole we see from the top) is also real. While instead, in the gravitational case the "hidden" pole is imaginary (it does not take real form). We shall revert to elaborate further on this point later, in relation to figure 5-8.

The above conditions can be compared with a conventional representation of a wave, which has "nodes" and "antinodes". The nodes correspond to the "fixed" points, and the "antinodes" to the curved part, as shown in figure 5-3, left part. It similarly holds for the waviness of a gravitational field, as it was also shown (simplified for schematic purposes) in the representation of the time curve in figure 1-1, where the waviness has nodes at points d′ and b′, and in-between these nodes there lies a positive antinode (the classical world we live in).

However, if we consider that the actual gravitational field is radial-like along 3 dimensions, the corresponding waviness is also extending in 3 dimensions. As shown at the right side of the same figure, that refers to having the one node (nodal surface) of the waviness upon the particle, and the other node (nodal surface) having the form of an immense event horizon lying at the very distant outer space (resembling a form of huge cosmic sphere). In other words, it's like the one node of the particle's gravitational field being at the particle itself (a type of "singularity"), and the other node being at the end of all observable universe (a "de-singularity"). This however, is only a simplified model, to be refined later.

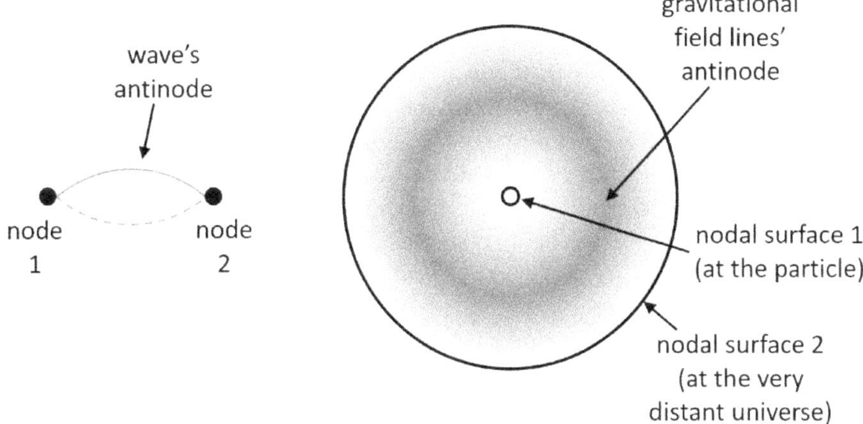

Figure 5-3: Antinode of a string (left) vs.
an antinode of a radial-like field of an elementary particle (right)

A more careful observation of figure 1-1 allows to consider that the nodal surface upon a particle (which supposedly resides near the center of our galaxy) does not refer to point b′ (where the time metric has zero value), but rather to point b, since a space inversion takes place there. Likewise, the other node (at the very distant outer space) might not have its gravitational field line vanish at point d′ (where the time metric has zero value again). It may be considered to rather extend to point d, where the *space* metric becomes nodal, even though the last portion refers to negative time. According to material to follow in section 6, these marginal regions b − b′ and d − d′ of a gravitational field "line" turn out to be of significance, as they relate to a particle's ability to **engage in effects of reaction** (like inertia or vortex flow). They also relate to an electron's ability to interact with neutrinos through the weak nuclear interaction (to be considered in section 9), and engage in interactions involving antimatter (like pair production or annihilation).

Furthermore, the fact that the metric of time has a precise value in our celestial neighborhood (point S in figure 1-1), and that it similarly holds for the metric of space, has a central role on the landing of a collapsed wave function (to be addressed in section 7).

For reasons of simplicity, at this point we shall neglect what holds at the nodal regions of the field lines (where either metric changes sign). What the remaining picture translates to, is that every conventional (fermionic) particle which has mass, is not exactly a matter monopole. What we perceive of it rather concerns the "real" pole of a hypothetical dipole, instead. The gravitational field lines which "come out" of this pole, spread radially out and away toward the very distant outer space, as if a "**hypothetical antipole**" existed there, which is not point-line, instead it is spread over the distant universe. There, the gravitational field lines pass through a distant hypothetical horizon associating to point d′ of figure 1-1, and beyond that distant range the gravitational field lines become **negative** (meaning that they follow the curvature in the negative metric of time). Even further away there is still another type of larger horizon which corresponds to point d, where the *space* metric turns negative, and the gravitational field lines stop being considered real, and become imaginary with respect to the classical-world's point of reference. From that point and on, the imaginary portion of each field "line" keeps following the rest of a full hypothetical loop, through the hidden portion which has no real context, eventually reaching back to catch up to the particle (via the hidden field portion). Notice that even though this seems to be an immense loop, its spatial extend is irrelevant, since it concerns a loop in time.

This general model of the gravitational field lines reaching toward infinity and then vanishing into the negative space metric may actually be subject to an improvement/refinement. In particular, while at point d′ the gravitational field lines "dive" into negative metric of time, a reference to figure 1-1 helps realize that the nodal point d′ corresponds to a horizon surrounding (lying outside) celestial formations of opposite energy, like ones that should be composed of antimatter particles. And, considering that a large number of celestial objects of such kind should exist throughout all distance universe[7],[8], we could assume that the far end of each gravitational field "line" of a conventional matter particle could eventually "dive" into such a distant celestial object. And through such diving, each gravitational field "line" would be able to eventually complete the hypothetical loop path. (Such a model also complies to the notion of the cosmos having a total energy balance equal to zero). Note that the above considerations concern only the four leptons (the electron, the positron, the neutrino, and the antineutrino) and not quasiparticles like the quarks (constituents of nuclear particles) for which conditions are somewhat different, and shall be addressed in section 10.

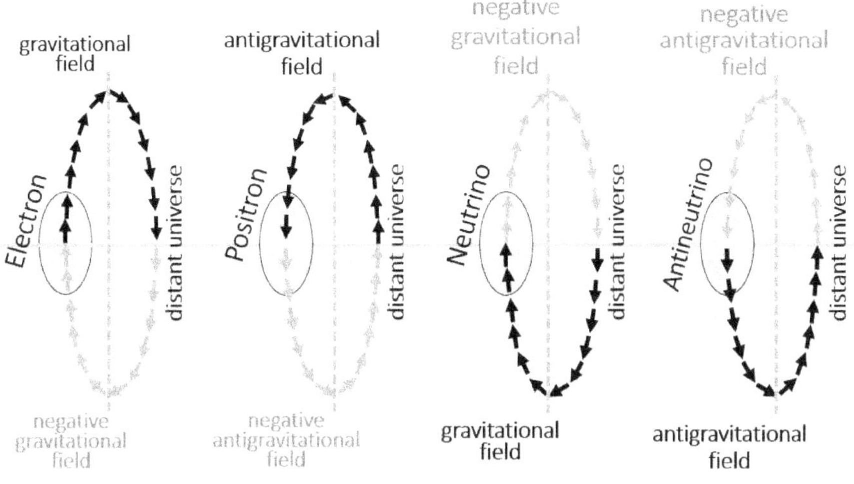

Figure 5-4: Real and imaginary parts of a gravitational field line of the electron and the other leptons, as seen from a hypothetical "side-view"

Figure 5-4 uses a similar schematic concept to that of figure 5-2 left part, to now illustrate what holds for a single gravitational field "line" of each of the four leptons, showing it from a hypothetical "side-view" which graphically depicts both the perceivable portion (upper half) and non-perceivable imaginary portion (lower half). Note that the elliptical shape of each particle is only schematic, it does not represent in any way the physical size or shape of the particle. Furthermore, since a gravitational field is radial-like, each field line is supposed to reach toward the "distant universe" where it crosses another node. Notice a tricky point, that the node at the "distant universe" resembles another hypothetical pole (other than the pole of the electron). Hence, out of a whole gravitational field line (constituting a hypothetical loop), <u>only the part to left to the vertical dashed line</u> relates to the electron, while the part to the right of the dashed line ascribes to an "other pole at the distant universe". In accordance to theory to follow, such "other pole" could refer to an antimatter particle lying at a distant celestial formation, but that point shall be elaborated ahead. Remember also, that for reasons of simplicity we have neglected what holds at the marginal portions of the field lines next to the nodes.

For the case of an electron (first on the left of figure 5-4) we may perceive the part of the gravitational field line which is viewable from "above" (the black arrows in the upper half), where the "above" represents our real classical world. And we may not directly perceive the negative gravitational field (grey arrows, at lower half) since that is imaginary. Along these lines, we may speak of the existence of **two variants of the gravitational field**, the ordinary real gravitational field which concerns the field of a real positive matter monopole (black arrows), and the imaginary gravitational field (grey arrows at lower half) which concerns the field of an imaginary negative matter pole. Certain comments about these two field versions may have as follows:

Ordinary Gravitational field, denoted by the symbol g or g_{re} (signifying "real"): This is the ordinary gravitational field which we have just described. It is a field which causes an **action** force, its field lines are **radial**-like. Each field line represents the trace of change in the metric of **time** (time passing faster the further away from a particle like an electron, and after some range passing slower until reaching the very distant universe). That concerns the perceivable portion of matter's or antimatter's gravitational field (e.g of electrons and positrons, if we speak of leptons only). Notice that the particular direction of the "arrows" shown in the figure (pointing either clockwise or anticlockwise) is only schematic, just to reflect the opposite directions in time passing between a matter particle (first from left in figure 5-4) and antimatter particle (second from left). As described earlier, regardless the direction of time, the gravitational field causes an attractive force in both cases (of electrons and positrons), since the parameter of time is involved in the power of two.

Notice that the upper half of figure 5-4 reflects the gravitational field in conditions of positive energy, while the lower half of this figure reflects the gravitational field that the corresponding particles exhibit in a cosmic region of negative energy. Therefore, the same kind of gravitational field (as of the black arrows) that applies between electrons and positrons in our cosmic region, also applies between neutrinos and antineutrinos and gets them attract each other provided they reside within cosmic regions of negative space metric which constitutes their natural habitat (see *lower half* portions of the third-from-left lepton and last-at-right lepton of figure 5-4) while that part is imaginary with respect to our real classical world.

Negative gravitational field, denoted by the symbol $\bar{g}$ or g_{im} (for "imaginary"): This is the gravitational field that neutrinos and antineutrinos exhibit toward our own cosmic region of positive space, as represented by the grey arrows (*upper half* portion of the third-from-left and the last-at-right leptons of figure 5-4), where the grey color is meant to reflect its imaginary (non-real) attributes with respect to our classical world.

A point of significance, is that the same kind of field that neutrinos and antineutrinos exhibit toward our world, also constitutes an imaginary gravitational field version of electrons and positrons (grey arrows at the *lower half* portion of the left two leptons of figure 5-4). While that field version is imaginary, so it does not take actual form, we shall consider it a "hidden" field, instead of a "non-existing" field, the reason being that under certain conditions it may engage in real interactions. In particular, as we shall consider ahead, this field portion engages in the real world as a **reaction field** in transitional cases which engage a "displacement-field" variant of it. For instance, in affecting vortex flow in fluids, as we shall consider later. The engagement of this field version in the real world is made possible due to the existence of the phase shift between the space and time metrics. In addition, this field version seems to also engage in the weak interaction, as we shall consider in section 9.

Inductive-like and Capacitive-like effects:

Just like the effects of induction and capacitance in electronics take place in association to a phase shift between the voltage and current alternations, it similarly holds in mechanics. In particular, in mechanics the phase shift develops in *space* (as per the phase shift between the Space and Time curvatures illustrated in figure 1-1), unlike in electronics where the phase shift develops in *time* (between the current and voltage alternations). The existence of this shift between the Space and Time metrics allows for each of g and $\bar{g}$ fields to interact with, and leak energy toward the opposite energy domain in each case during transitional cases, through a Sine or a Cosine coordinate.

The possibilities that arise in mechanics concern either an "inductive-like" phase shift in space (a drag), or an "capacitive-like" phase shift in space (a lead). An introduction on corresponding interactions may be outlined as follows, while more related material shall be provided in section 6.

Inductive-like behavior:
This involves a version of the gravitational field which applies in a symmetric way to as the electromotive potential applies in self-inductance in electronics. This field version turns out to be responsible for the generation of the **inertial** impedance. It is a field of negative energy, as it concerns a field coordinate projecting toward the imaginary axis. Due to its imaginary character, it does not produce acceleration by its own, instead it impedes a change in velocity, so it operates as a **reaction** field, and consumes energy without doing work. For reasons of symmetry in the naming with the electromotive potential, we shall name the corresponding mechanical potential "**gravitomotive**", and we

shall describe it through the term $\oint g \cdot d\lambda$ (in analogy to the electromotive potential which is described by the relation $\oint E \cdot d\ell = -L\, dI/dt$). The $\oint g$ is meant to concern the imaginary version of the gravitational field, which has loop-like field lines, just like $\oint E$ is considered to have loop-like field lines in electronics since it applies over a circuit loop. The term $d\lambda$ concerns a loop in *time*, which rides the axis of XX' of figure 1-1. As mentioned earlier, even though this loop may seem immense in spatial terms, its spatial extend is irrelevant as it concerns a loop in *time*). In relation to material to be considered ahead, the field version involved in the gravitomotive effect is a displacement-field, and we shall denote it by the term $\bar{g}_{d-}$, where g indicates the gravitational field, the upper bar signifies its imaginary field version, the subscript "d" indicates its displacement-field character, and the negative sign "−" indicates its inductive-like (instead of capacitive-like) essence. The energy of the gravitomotive potential may be accounted for through a **sine** coordinate, which corresponds to a component which misses to deliver actual work in the real domain (while it consumes energy), and applies as an imaginary coordinate in symmetry to the electromotive case.

An orthogonal embodiment of the gravitomotive potential $\oint g \cdot d\lambda$ is also possible, where that may generate reaction motion of its own. Due to its negative-energy nature, its field lines project to the real world through an inverse geometry (in association to an effect of displacement that will be stated to apply for imaginary field coordinates in general) and this gets it materialize over a loop-like trail. This potential is responsible for the creation of **vortex flow** (as in the whirling of water), which corresponds to a mechanical counterpart of the inductive current in a metal ring when there is a change in magnetic flux through the ring as per Lenz's law $\oint E \cdot d\ell = d\Phi_B / dt$. By symmetry, the corresponding mechanical potential $\oint g$ has the freedom to displace matter in space, over a loop which deploys in time $d\lambda$, affecting a "displacement momentum". This loop $d\lambda$, and corresponding displacement momentum upon this loop, drives the vortex flow. This embobiment should involve the **cosine** coordinate (a "real" coordinate) of the "displacement gravitational field" $\bar{g}_{d-}$. We shall later figure that in the actual case a particular inversion takes place, and the interaction actually gets to engage a *sine* (imaginary) coordinate of a "displacement **gravitomagnetic** field" which has imaginary character (to be introduced and elaborated later), instead of the mentioned *cosine* coordinate of the displacement **gravitational** field. This point requires certain theory that shall be presented ahead. According to it, this interaction gets to leak energy from the imaginary domain toward the real domain, which causes the vortex flow.

<u>Capacitive-like behavior:</u>
An analogous process may apply in the opposite direction in the case of a capacitive-like phase shift between the curvatures of space and time. This may generate a "capacitive" type of "displacement motion". In particular, just like the electric field may come in a "displacement electric" field version $D=\varepsilon E$, it similarly holds for the gravitational field which may come in a "**displacement gravitational**" field version which we shall denote with the symbol g_d where the subscript "$_d$" stands for "displacement". This refers to a relation of the form

$$g_d = \varepsilon_g\, g \qquad\qquad\qquad (5\text{-}1)$$

where the term ε_g corresponds to a permittivity-like term which will be elaborated later, and we may call it mechano-permittivity. The subscript "$_g$" stands for "gravitational" and is meant to differentiate the mechano-permittivity from the electric permittivity ε. In the gravitational interaction, the term ε_g relates to a capacitive-like phase shift, and concerns a shift in *space* (instead of a shift in *time* of the electric case) between the Space and Time curvatures. The ε_g is typically a complex entity (involving a Real and an Imaginary part), just like it holds for its electric counterpart. Consequently, the displacement gravitational field is more accurately described through a relation like

$$g_d{+} = \varepsilon_g\,(g\,\cos\theta - i\,g\,\sin\theta) \qquad\qquad\qquad (5\text{-}2)$$

Actually, since the displacement here is of capacitive nature (instead of inductive which was the case above) we have now added a "+" sign next to g_d, to indicate a shift forward-in-space (instead of backward). Once again, the relation (5-2) is symmetric to the displacement electric field's expression of $D=\varepsilon\,(E\cos\theta - i\,E\,\sin\theta)$. The fact that the $g_d{+}$ has two coordinates (a cosine and a sine) allows it to also become involved in two orthogonal interactions, as follows:

The imaginary version of the capacitive displacement gravitational field may apply its energy through a **sine** coordinate $\bar{g}_d{+} = \varepsilon_g\,(- i\,g\,\sin\theta)$, which may be alternatively expressed in the form $\bar{g}_d{+} = -\varepsilon_g\,(\bar{g}\,\sin\theta)$, where the imaginary term i indicates that it concerns the negative energy version of the gravitational field. (To clarify, the negative energy version of the field stems from the existence of i, and not from the negative sign involved). Despite its imaginary essence, this version of the field may however bring real outcome in the real world, by interacting with the imaginary part of a particle's wave function, as the two imaginary parts produce a negative real result. Actually, in this case the negative real result together with the negative sign of above, yield a positive outcome. The corresponding displacement produced, is reflected as a shift forward in space of the particle's wave function, and that corresponds to a displacement of momentum p_d (where the subscript "$_d$" stands for displacement). Since the process concerns a shift of waviness, it associates to the particle's wave nature (not its particle nature). An embodiment of such an interaction concerns the effect of **barrier penetration** of quantum mechanics, which turns out to be symmetric to the electric effect of the displacement of electrons across a capacitor's plates in the form of displacement current I_d. As the interaction involves the negative-energy part (imaginary part) of the particle's wave function, this is accounted for as a "displacement of gravitational flux", instead of conventional motion, and this is symmetric to the process of displacement of electric flux through the plates of a capacitor (instead of conventional motion of charge), as will be analyzed in section 6. The effect of barrier penetration turns out to correspond to an effect of tunneling forward in *space*, in symmetry to its electronic counterpart of the displacement current taking place between the plates of a capacitor, which corresponds to a tunneling forward in *time* (through a capacitive phase shift). Furthermore, as it corresponds to an effect of wave-nature (involving the imaginary part of the particle's wave function), the displacement of the particle's matter wave is analogous to what holds in frustrated total internal reflection of waves. In this particular case, the process **mediates the transfer of displacement momentum p_d**. Because of the imaginary essence of the interaction, the corresponding displacement field becomes evanescent in short range over the real world. This is why the penetration gets significantly reduced by an increase in the width of the barrier. Notice also, that the same field variant shall also be explained to become involved in the weak interaction.

An orthogonal version of the capacitive displacement field version is also possible, which should involve opposite sign of energy, and an associated inverse geometry. Due to the orthogonality, it should be accounted for through a **cosine** coordinate of the displacement gravitational field $g_d{+} = \varepsilon_g (g \cos\theta)$. Here again (similar to as in the inductive version of above), a particular inversion takes place, due to an effect of reciprocity. Due to this inversion, the interaction actually gets to engage a *sine* (imaginary) coordinate of a "displacement **gravitomagnetic** field" (to be introduced and elaborated later), instead of the mentioned *cosine* coordinate of the displacement **gravitational** field. In relation to material to follow, this interaction involves a *shift* in space Δx which applies in the form of a capacitive-like potential, having radial-like field lines (instead of loop-like). An embodiment where that potential becomes involved, is in powering "exergonic" biochemical reactions, which release more energy than they absorb. This is possible as the interaction leaks energy from the imaginary domain toward the real domain. As we shall see ahead, that is what drives life-powering interactions.

5.2 The Electric Field (E)

The electric and the gravitational interactions seem to be symmetric to each other in various ways, like in the form of the relations governing their corresponding forces ($F_E=qE$ and $F_g=mg$). Or, in that they both seem to form monopoles which have radial-like field lines. Actually, provided that the gravitational field of a particle corresponds to a curvature of time, a somewhat similar condition should be expected to hold for the electric field of a particle, involving some analogous type of (spacetime-related) curvature.

Regarding points where the gravitational and electric fields *differ* from each other, this refers to indications that we shall be addressing ahead, that electrism and gravity operate in complementary grounds to each other through a particular exchange between parameters of Space and the Time. One area where that may be traced, is in effects of reaction where -for instance- in electronics the reaction shifts deploy *in time* (like in the case of a capacitive or an inductive phase shift in time between voltage-current oscillations in an electric circuit), while in mechanics the reaction phase shifts deploy *in space* (as per the shift between the Space and Time metrics shown in figure 1-1).

Such a symmetry and complementarity appears to have deep fundamental grounds, as we shall soon figure that the difference in essence between the electric and the gravitational fields and interactions appears to be subject to a particular swapping between space (actually negative-reciprocal space) and time. That apperas to get the electric and gravitational interactions associate to each other in an orthogonal way, meaning that attributes of either one of the two (gravitation or electrism) are imaginary with respect to the other. This is why the two interactions do not seem to interact between each other, at least in a first level of consideration.

A question that arises in relation to the above preliminary pressumption is that, if either of these two interactions is considered orthogonal with respect to the other (therefore either one being imaginary with respect to the other), then how can it be possible that *both* the electric *and* the gravitational interactions are perceived in the real classical world around us. Or, which one of these two should be considered of "real" origin, and which one shouldn't, but may still project to the classical world. And for the one that shouldn't, a question then arises as of how it may get to project to the real world having the form of an ordinary classical interaction. It turns out that each one of these two interactions may come up in two variants, a real one and an imaginary one. This is so, since either one may relate to energy of positive or of negative sign, where effects in positive energy should be real and effects in negative should behave as imaginary. At this point we shall introduce a particular concept, that shall become more clear as we move on, that **space (that is, "positive space") is inherently imaginary (orthogonal) with respect to time**. Actually, if this is so, then the "negative-reciprocal space" (which bears characteristics of "imaginary space") should behave as "imaginary of the imaginary" where that yields a "real outcome of negative sign" due to the relation $i^2=-1$. In such case, it seems to turn out that **a difference in the metric of "negative-reciprocal space" refers to the essence of the electric field** (in analogy to as a difference in the metric of time identifies to the gravitational field). In fact, the negation involved in the term $i^2=-1$ seems to be responsible for the electric force to be heading in opposite direction to that of the gravitational force, in the sense that two charges of same sign *repel* each other, while two particles of same sign of matter (conventional matter) *attract* each other. Actually, we need to point out that the question of what is real and what is imaginary (acting as negative-reciprocal) may depend on the frame of reference (like for instance in which particular cosmic region of figure 1-1 does an interaction take place). In our own cosmic region, the physical representation of "curvature in negative-reciprocal space" appears to match to the essence of the electric field.

The above consideration that the "imaginary of the imaginary" yields a real physical outcome obviously finds a match in numerous physical interactions, aside the electric one. This includes non-classical interactions (engaging imaginary field versions) which get to exhibit real results upon the real world due to the co-involvement of two imaginary entities. For any and all such cases, it is essential to explore how such a projection to the real world should be looking-and-behaving like. For figuring that out, we shall re-write the relation $i^2=-1$ in the form $i=-1/i$. The latter expression is of significance, as it allows to realize that there may exist properties and effects happening all around us, which we are accustomed to deal with in our classical world, but these are actually of imaginary origin, and get to project upon the real world only through such process of negative reciprocation, without that having been realized. In relation to that, as we move ahead we shall bring to attention imaginary properties interacting "with respect to reciprocal space $-1/dx$" (instead of "with respect to space dx"), or likewise "with respect to reciprocal time $-1/dt$" (instead of "with respect to time dt").

This feat will allow us to resolve numerous counterintuitive issues in physics, like for instance the units of the Planck constant h being Joules•seconds where the parameter of time is not in a denominator (to be addressed in section 7). Or, that the units of the electric *flux* $\Phi_E = E\,dx^2$ have the parameter of area dx^2 in numerator position, instead of denominator. That shall allow us to obtain a far deeper understanding of symmetries between established equations. In one example, if the expression that describes the propagation of a travelling electromagnetic wave $dE/dx=dB/dt$ is re-written in the mathematically equivalent form $dE/(-1/dt)=dB/(-1/dx)$ where the electric field changes with respect to negative reciprocal (therefore imaginary) time, and the magnetic field changes with respect to negative reciprocal (therefore imaginary) space, this may provide an indication that Lenz's law concerns an opposite-energy variant of Faraday's law (to be addressed in section 5.3).

Moreover, if in some other circumstance we tried to *impose* the substitution of the parameter dx by $(-1/dx)$ and the substitution of dt by $(-1/dt)$, where obviously that result would NOT be mathematically equivalent, then such negative reciprocation of both space and time would lead into an effective **swapping** between the parameters of space and time. If we try such a change in a velocity term, and use reciprocals of the space and time $(-1/dx)/(-1/dt)$ instead of the conventional (dx/dt), that makes for **dt/dx**. In that case the swapping between the numerator and the denominator yields the "reciprocal of classical velocity, $1/v$". And since the electric and gravitational interactions seem to involve an exchange of (both) space and time with respect to each other, we shall keep that in mind to soon see (in section 5.4) how such exchange of parameters differentiates the electromagnetic interaction from the gravitational interaction.

Notice that this feat of reciprocity, along with a corresponding exchange of the parameters of Space and Time, complies to the aforementioned assertion that space is inherently imaginary (orthogonal) with respect to time. Actually, we may now extend that by considering that **effects involving either one of them (space or time) is being counterbalanced by effects involving the negative reciprocal of the other**. Alongside that concept, since the gravitational field of a particle has been stated to correspond to a curvature in the metric of time, the **electric field seems to co-exist as a curvature in the metric of negative-reciprocal (therefore imaginary) space**. Actually, without contravening the conventional assertion that space (positive space) goes side-by-side with time (positive time) in what is conventionally considered as "spacetime", one may accept that since negative-reciprocal space is orthogonal to real positive space (in the sense that either one is imaginary with respect to the other as they should be represented in perpendicular axes), the existence of an electric field (the curvature of negative-reciprocal space) is orthogonal to (and therefore does not interact with) the gravitational field (the curvature of positive time). That is, these two fields may concurrently exist and act as independent fields.

In accordance to material that will be considered in section 9 and refers to energy balancing, the conventional notion that time and space are tightly interrelated entities, requires certain enhancement. That is, the curvature of time (which concerns a particle's gravitational field) should be considered to go side-by-side with the curvature of **negative-reciprocal** space (like the curvature concerning a particle's electric field). In fact, the co-existence of these two curvatures gets to balance out their energies (to null), so the co-existence of these two fields is favored by nature.

In terms of raw **units** of the electric field, and in accord to what will be presented in the next section concerning the magnetic field, it arises that the electric field's raw units should correspond to an "inverse-acceleration-like" expression, of units sec/meter^2 (or $dt/d(-x)^2$, where $-x$ concerns negative space). Note that this inaccurately represents the curvature of negative-reciprocal space, much like gravity's units dx/dt^2 inaccurately represent the curvature of the metric of time, yet both work well in calculations due to the principle of equivalence. In fact, this sort of symmetry (and complementarity) between the electric and gravitational fields will later be shown to extend to a similar symmetry and complementarity between charge and matter, as will be introduced in sections 7 and 8. That will actually come to justify why charge and matter co-exist in elementary particles like the electron (or the positron), since the energy of the one balances-out the energy in the other. And likewise, it similarly holds for more exotic fermions which will be explained to carry "anti-charge" (to be associated to weak interaction's "neutral charge") and "negative-matter".

Furthermore, we shall later figure that the curvatures associated with both the gravitational and electric fields should not be understood as being a "consequence to the environment" due to the existence of matter or charge. Instead, a particle's fields constitute an integral part of that particle's essence. Furthermore, as it was explained to hold for the case of the gravitational field, the electric field lines also follow a waviness as the one illustrated in the curves of figure 1-1, but now concerning negative-reciprocal space. This means that the corresponding field lines should incorporate a difference in the metric of negative-reciprocal space which increases until a certain point at distant outer space, then takes an opposite slope toward classical infinity, until a point where at the field lines reach a nodal surface and then reverse sign (therefore turn to be of imaginary essence). So, what we perceive of the electric field is only the radial-like real portion of the total curvature involved, while the nodal surface where sign changes corresponds to the surface/horizon of celestial formations like quasars and blazars. As a large number of such celestial objects exist in nature (spread away per every direction over the very distant outer space), their surface applies as the "other end" of the real portion of each electric field line. In this sense, such distant celestial formations apply as **"poles of anti-electric field"**, whose corresponding "anticharge" is imaginary with respect to us, so these celestial objects are not considered "charged" in the conventional sense. Their "anticharge" comes along with an "antielectric field" outside these celestial objects (point d of figure 1-1, as viewed from its right side).

Coming back to the notion that the "other end" of the electric field lines of conventional charges reach to the surface of celestial formations like quasars and blazars, we need to figure what would happen then. Beyond that range, the electric field lines should become imaginary and hidden from classical perception, and carry out the rest of a hypothetical full loop through the imaginary side (without having real form) until they eventually reach back to catch up and re-appear at the real electric **"charge monopole"** of the electron. (More will be said about charge and anticharge poles in section 5.3). Furthermore, while the charge of an elementary particle does arise directly from the Schrödinger equation, its nature could be approximated through a "complementary-to-the-Schrödinger" equation, to be introduced in section 8.

Regarding the way that the electric force applies and gets charges repel or attract each other, this takes place under the same general concept of "releasing potential energy" as was described earlier for the case of the gravitational interaction, subject to a swapping between the parameters of time and negative-reciprocal space. Namely, when a negative charge is subject to the electric field of another negative charge and accelerates, its acceleration causes a similar Doppler-like differential in the stretching of negative-reciprocal space, with that being LESS contracted toward the direction of acceleration. This Doopler-like differential develops in the opposite direction to the differential attributed to the essence of the (externally applying) electric field, having the negative Space being MORE contracted toward the particle. In such case, when two charges accelerate away (or toward) each other due to an electric repulsion (or attraction), these two differentials counterbalance each other, and the potential energy in the curvature of the negative-reciprocal space metric (the electric field) gets released. This release is what allows and drives for the acceleration to develop by its own. In the case of two charges of same sign being close to each other, the direction of the force points in the opposite direction to that of the force of gravity, this arising as per the opposite energy between the two interactions, as previously explained through the negation in $i^2 = -1$.

In a point of attention, according to existing theory two charges "communicate" to each other their proximity (so as to exert an electric force to each other) by the exchange of virtual photons. Hence, at the velocity of light. This approach, however, does not provide a full picture of the interaction, so it is not fully accurate. It turns out that an electric field may indeed communicate (deploy) its presence in a region at the speed of light. But once its presence is established and the electric field is there, then effects of reaction which involve a phase shift, may allow for an unconventional means of transmission: In relation to material that will be considered later, such effects of reaction concern negative energy, exhibit imaginary attributes, and allow for mediation at superluminal fastness. This actually address the so-called **"spooky action at**

a distance". It now appears that this is not spooky at all. A minimal change drives a small **reactive phase shift** in time or in space, and that phase shift is conveyed loop-like at once, just like all elements of a bicycle's wheel rotate all at once. Such a condition may be met in the case of entanglement, while another case concerns an invention made by the author (not included in this book) regarding a new type of modulation for **ultrafast transfer of energy**.

Another point of importance, concerns the nature of **positive** charge at the *elementary particle level* (for example, the charge of positrons). Note a distinction, that at the elementary particle scale, the charge is fundamental, while at the macroscopic scale the positive charge could be only relative, like for instance when a pole is considered to be "positive" because it has a lesser number of negative charges (electrons) with respect to another pole. At the elementary particle level we may consider "ordinary" the negative charge only, like the charge of an electron, whose sign could be attributed -as previously mentioned- to the negation in $i^2=-1$. The positive charge, in the other hand, would *not* be considered fully ordinary at the elementary particle level: Recall that the positively charged leptons, the positrons, are not stable particles, since they are vulnerable to annihilation. The stable positive charge found in atoms comes from the protons, which are stable particles, however the protons are not truly elementary particles, as they are made of quarks where different conditions apply and allow to keep these composite particles stable. In fact, even though the positive charge of these two sources (a positron and a proton) may seem equivalent in terms of charge value, the nature of charge in either of these two particles has different origin and behavior. In particular:

- The positron is a truly elementary particle, it lives in the negative time metric, and its electric field corresponds to raw units of $(- sec)/(- meters)^2$, which makes for $(- sec)/(meters)^2$, where the negative sign of time (in the numerator) may be assumed to reflect that the electric force exerted by a positive charge is heading in the opposite direction than the electric force exerted by a negative charge. Since positrons live in negative time (negative energy in time), positrons are not stable particles in our cosmic region, they are vulnerable to interact with electrons of same energy in positive time and mutually engage in an annihilation event.
- The positive charge of the proton, in the other hand, has a differentiated origin. The proton is not an antiparticle, so it does not exactly "live" solely in the negative time metric. Its charge corresponds to the sum of the charge of its constituent sub-particles, the quarks. As will be presented in section 10, each of the quarks engages into both negative and positive energy in time concurrently, due to the involvement of a particular phase shift that will be explained to constitute the differentiation of quarks from leptons. This gets quarks to partially inhabit both the positive and negative energy

domains in parallel, and that actually protects them from annihilating via interaction with particles that live solely in positive energy of time. Furthermore, the quarks match with each other so as to satisfy quantization criteria collectively, and in this why they may reach stable form, jointing together through the strong force. The charge of the composite particles that quarks form, corresponds to the sum of their fractional charges.

That the quarks engage into both positive and negative energy seems to also provide to them an additional differentiation. In relation to material that will be considered in section 10, the sourcing of quarks' positive charge (where positive charge concerns energy in negative time) along with the co-involvement of the negative space curvature, gets **the product of these two negative metrics yield a positive energy value**, as referred to for region b′ – d of figure 4-2 lower part. This seems to constitute an important difference in the way the corresponding positive charge becomes involved in the creation of a potential. In particular, it seems plausible to exhibit enhancing characteristics (the opposite of impeding). An embodiment of this behavior seems to concern the potential developed between the poles of an electric battery, which gets to provide long-lasting power. This is unlike what holds between the plates of a capacitor which rely on leptonic potential difference attributed to shortage of electrons in one of the plates with respect to the other, which may discharge quickly.

In our own cosmic region there seems to be an almost equal abundance of negative and positive charge. Even though in our immediate environment there may exist individual macroscopic objects which are negatively charged or positively charged, or, at a somewhat larger scale there may be meteorological effects creating spots of a particular sign of charge, at the celestial body scale there seems to be an almost equal total amount of positive and negative charge contained, and this lets conventional celestial bodies NOT exhibit a net electric field. (This of course is unlike what holds for the gravitational field, where a net gravitational field arises since all matter in our region is of "positive sign of matter"). And while gravity identifies to the curvature of the positive time metric, the corresponding curvature of the positive *space* around celestial formations does not associate to an electric field, as the electric field seems to correspond to a curvature in the *negative-reciprocal* space metric.

In a related side-note, since the curvatures of positive time and positive space are subject to a phase shift between them, this implies that these two behave as distinct entities, and this actually comes in support to the notion that the essence of the gravitational field should not concern the curvature of *both* time and space metrics. Provided that gravity has been earlier explained to associate to the curvature of the metric of time (only), the curvature of the positive space

metric should associate to a different field. We shall later associate it to the
"antielectric field", and shall further describe a particular relation that has with
neutrons in atom's nuclei. Furthermore, other than the "*curvature* of space",
later we shall also address a (distict) concept of a "phase *shift* in space".

As per all the above, figure 5-5 illustrates schematically a single electric field
line of each one of the four leptons (electron, positron, neutrino, antineutrino)
from a hypothetical "side view", in a similar way to as was done in figure 5-4
for the gravitational field. For instance, for the electron (first from left) we may
realize the perceivable portion of a particle's negative electric field line (black
arrows, upper half) as this is viewable from "above", where the "above"
represents our real classical world. And we may not directly realize the non-
perceivable portion referring to an anti-electric field (grey arrows, lower half)
since it is imaginary (and therefore hidden to our classical world, as we look at
it from above).

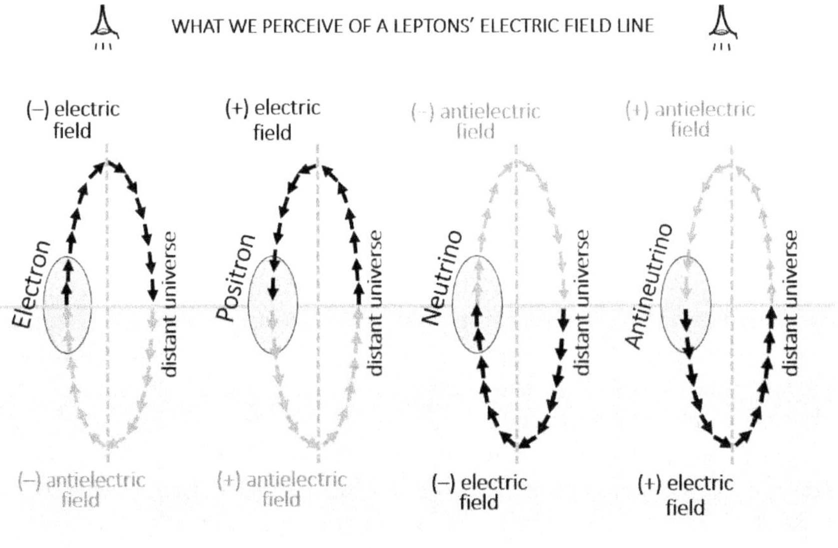

<u>Figure 5-5</u>: Visible and non-visible portions of an electric field line of
an electron and other leptons, viewed from a hypothetical "side view"

Note the tricky point again, that the node at the "distant universe" resembles
another hypothetical pole (other than the electron's pole). Hence, from the
whole electric field line, <u>only the portion to the left of the vertical dashed line</u>
relates to the electron, while the portion to the right of it ascribes to the "other

pole at the distant universe". That means that the value of charge of an electron Q=-1e ascribes to the electric field to the left of the dashed line (and we shall come back to this later). Remember also, that for reasons of simplicity in this figure we have neglected what holds at the marginal portions of the field lines next to the two nodes (at the particle, and at the "distant universe").

In relation to what that figure illustrates, we may account for the existence of two basic variants of the electric field, as follows:

Ordinary Electric field (E or E_{re}):
This concerns the ordinary electric field which we have just described (re. figure 5-5, upper half, for the left two leptons). It is a field which causes an **action** force, and its field "lines" are **radial**-like. As discussed, each field "line" represents the trace of change in a differential in the metric of **negative space**. Note however, that this applies with respect to the reference point of classical world of mechanics. If considered with respect to the reference point of electronics (how other charges perceive it, then each field "line" should be considered to represent the curvature of real space (due to the orthogonality between the gravitational and electric interactions, further elaborated ahead). In any case, the direction of the field "arrows" is only schematic and represents the opposite directions of time for the electrons and positrons respectively.

Note that the same type of field and potential is involved in the interaction between neutrinos as well as antineutrinos *among themselves*, and gets them force each other **provided** they reside within cosmic regions of negative space metric which constitutes their natural habitat (and is imaginary to us), re. *lower half* portion of the third-from-left and the last-at-right leptons of figure 5-5. Note however, that what this figure illustrates schematically, does not apply for the charge of quarks, which shall be considered later, in section 10.

Anti-electric field ($\bar{E}$ or E_{im}, or descriptively as $\oint E$ as it deploys loop-like):
The antielectric field constitutes the field that neutrinos and antineutrinos exhibit toward our own cosmic region. It is represented by grey arrows (figure 5-5, *upper half* of third lepton from left, and last at right), where the grey color is meant to reflect its imaginary attribute with respect to our classical world.

A point of significance, is that the same kind of field that neutrinos and antineutrinos exhibit toward our classical world, is of same essence to the imaginary field variant of electrons and positrons (grey arrows at the *lower half* portion of the left two leptons of figure 5-5). While this field is imaginary, we shall consider it a "hidden" field (and not a "non-existing" field), the reason being that under certain conditions it may engage in real interactions. In particular, as it will be considered ahead, this field portion engages in the real world as a reaction field in transitional cases, like for instance in affecting the electromotive potential that causes a displacement current in a metal ring when there is a change in magnetic flux through the ring (as per Lenz's law). That this engages in the real world, is feasible thanks to an (inductive) phase shift between the space and time metrics. Later, in section 9, we shall also consider how this field variant seems to also engage in the weak interaction.

In terms of units, both these variants of the electric field may be considered to have same units (sec/meter^2), however the antielectric variant comes along with an imaginary term i. When this imaginary field variant interacts with another imaginary entity, that brings a negative *real* result as per $i^2 = -1$. In that case, **the projection toward the real world may be accounted through a reciprocation of parameters, since it holds $i = -1/i$, and by affecting such reciprocation the imaginary term may get waived**. In that case, the units involved could be treated as $(1/\text{sec})^2/(1/\text{-meter})^2$ which equals **meter^2/sec**. The reciprocation goes along with an inverse geometry, with that version of the electric field having loop-like field lines (instead of radial-like). In relation to material to follow, **any such loop-like shaped (reaction) field should be considered to counterbalance an orthogonal radial (action) field/potential**.

Inductive and Capacitive effects:
A phase shift between the voltage and current alternations associates to effects of inductance and/or capacitance. These concern a shift backward or forward in *time*. (This is in analogy to a shift in *space*, shown in the illustration of figure 1-1, which shall be explained to resolve numerous effects of mechanics). The existence of such a shift allows for each one of the E and $\bar{E}$ fields to interact with, and leak energy toward the opposite energy domain in each case, during transitional interactions, where energy is provided through a Sine or a Cosine coordinate correspondingly.

- Effects involving an inductive phase shift in time (a lag) include self-induction, or the creation of a current in a metal ring when there is a change in flux through the ring. These two concern orthogonal settings, they both involve the electromotive potential, and leak energy toward the opposite energy domain in each case (to be elaborated ahead).
- Effects involving a capacitive phase shift in time (a lead) include the displacement of current between the plates of a capacitor, as well as a similar process which concerns the powering of electric batteries. These are orthogonal to each other, engaging the magnetic counterpart of the electromotive potential, which we shall call magneto-motive.

The above effects will be described in section 6.

5.3 The Magnetic Field (B)

The magnetic field is described through either of two, separate, but closely related representations, B and H, which associate to each other through the relation $H=(1/\mu_o)B$, where μ_o is the magnetic permeability, a measure of the magnetization that a material obtains when it is subject to an applied magnetic field. The B refers to the **magnetic flux density** and is measured in units of Tesla, while the H refers to the **magnetic strength** or the **auxiliary magnetic field** and is measured in Amperes per meter. According to textbook theory these two representations differ on how they account for magnetization, while in vacuum they are the same. However, according to the present theory, these two expressions differ from each other in a fundamental way, as H involves the magnetic permeability μ, where that has a deeper role than simply being a measure of magnetization of materials. This is so, as μ is measured in Henries per meter, so it involves inductance L, where L is typically a complex entity (involving a real as well as an imaginary part) and directly associates to the presence of a phase shift between the Voltage and Current oscillations in electric circuits. That makes H far richer in terms of describing the possible ways that the magnetic field may interact. Moreover, since the expression of H involves the term $(1/\mu_o)$, it acquires the character of a **"displacement magnetic"** field (much like its electric counterpart $D=\varepsilon E$ corresponds to a "displacement electric" field). Note that in the next paragraphs we shall be referring mostly to the B field (instead of H), due to that being a more raw, non-displacement field version, and wherever applicable we'll involve the permeability regulator μ. This way it shall be more clear how each element (B, μ, L) becomes involved in every interaction we will be considering.

The way to trace into the essence of the magnetic field is to look into the relations which describe its behavior, like Ampere's law and Maxwell's equations. According to the former (5-3), the magnetic field B arises when there is motion of charge (like a current I flowing through a circuit loop). According to (5-4), the magnetic field also arises in reaction to a change in the electric "flux" (not "field"), and more particularly a "displacement" electric flux $\varepsilon_o \Phi_E$, per time. Furthermore, according to (5-5), a change in the magnetic "flux" (not "field") per time $d\Phi_B/dt$ generates an electromotive potential $\oint E \cdot d\ell$, where the field mediating this potential is not the conventional electric field, since it has loop-like field lines (as it deploys over a circuit loop).

$$1/\mu_o \oint B \cdot d\ell = I \tag{5-3}$$

$$1/\mu_o \oint B \cdot d\ell = \varepsilon_o \, d\Phi_E/dt \tag{5-4}$$

$$\oint E \cdot d\ell = - d\Phi_B/dt \qquad\qquad\qquad (5\text{-}5)$$

We shall temporarily leave aside the relation (5-3) which describes the magnetic field around a current in a circuit loop, and we shall first focus in the other two relations which concern a *changing* or *alternating* field. One may notice that (5-4) and (5-5) are complementary to each other, in the sense that they involve a swapping of the parameters of space and time with respect to which each field changes. That is, in (5-4) the electric part changes per *time* (*dt*) and the magnetic reaction develops in *space* (*dℓ*), while in (5-5) the opposite holds, it's the magnetic part that changes per *time* (*dt*) and the electric reaction develops in *space* (*dℓ*). In relation to that, the following points of attention are raised:

Point 1: Counterbalancing roles of *E* and *B*, and identity of the *B* field

The fact that when the one field (either the magnetic or the electric) changes per TIME the other one reacts in SPACE, suggests that **these two fields have counter-balancing roles in opposite domains** (e.g. the time and the space domain). This can become more apparent if in (5-5) we substitute the magnetic flux (which concerns the flow of *B* through a surface area dA, or dx^2) with the relation $\Phi_B = B\ dx^2$, in which case (5-5) becomes

$$dE/dx = dB/dt \qquad\qquad\qquad (5\text{-}6)$$

This expression describes the propagation of a travelling electromagnetic wave (EMW). Actually, a simple re-arrangement of its terms as per $dE = (dx/dt)dB$ also refers to the well-known relation $E = vB$ which describes the relation between the electric and magnetic fields.

Despite the matching between these equations, the particular transformation from (5-5) to (5-6) is however loose, since these two equations apply under a different set of conditions. The former concens alternation of charge, while the latter concerns the propagation of fields of an an EMW, where the wave arises in response to alternation of charge, and deploys in an orthogonal direction. More specifically, the electric field in (5-5) deploys over a loop (the electric circuit's loop), and according to conventional theory it has loop-like field lines. We shall here assume that this concerns an *imaginary version* of the electric field, which is the version that mediates the electromotive potential. Likewise, the magnetic field in the same relation is involved in the form of FLUX Φ_B (instead of a plain field *B*), and we shall here translate this to indicate that (5-

5) concerns a measurement of the magnetic field with respect to the negative (imaginary) space metric. This is so, as the magnetic "flux" is measured "with respect to reciprocal space", since the term dx^2 is in the numerator, so it holds: $\Phi_B = B\ dx^2 = B\ /\ (1/dx^2)$, which describes the magnetic field per area of imaginary space $(1/dx)^2 = -idx^2$. These differences in the setting of fields between (5-5) and (5-6) has as a consequence that: (i) the latter relation describes a different orthogonality between B and E than what the former equation describes, and (ii) that may lead into certain new insights on how EMWs propagate, which we shall address shortly.

The relation (5-6) is significant as it reveals that a change in the one field per **space** is equivalent to a change in the other field per **time**. This may allow to make a supposition, which we shall be able to support later: That on the basis of symmetry, and, provided that the electric field is presumed to associate to a curvature in the imaginary (negative reciprocal) space metric, **the magnetic field should likewise associate to a curvature in the metric of imaginary (negative reciprocal) time**. Actually, this imaginary essence is key in differentiating the magnetic field, from the gravitational field which was stated to correspond to a curvature in the positive metric of time. It shall soon arise, that **the field lines of B correspond to equipotential trails, in terms of "fastness of passing" of imaginary (negative reciprocal) time** idt $(=-1/dt)$.

<u>Point 2</u>: The B field in EMWs, and EMW propagation

A closer look to the particulars of the propagation of EMWs may also allow to obtain a deeper understanding on the essence and behavior of the magnetic field. So far, we have referred to Lenz's law (5-5) and obtained (5-6) from it. We shall now turn attention to the Maxwell's equation (5-4) as this leads to another equation that concerns the propagation of EMWs, through a complementary (but jointly required) relation. Working as previously, and plugging into (5-4) the electric flux term $\Phi_E = E\ dx^2$, we obtain:

$$(1/\mu_o)\ dB/dx = (\varepsilon_o)\ dE/dt \qquad\qquad\qquad (5\text{-}7)$$

Once again, the transformation from (5-4) to (5-7) is loose, since these two equations apply under different conditions. The former concerns a change of the electric field along the direction of the field "arrows", while the latter concerns the propagation of an EMW which deploys in an orthogonal direction. Moreover, the electric field in (5-4) is involved in the form of a FLUX Φ_E (not a plain field E) which -once again- indicates that it is measured "with respect to the negative reciprocal" space metric $\Phi_E = E\ /\ (1/dx^2)$. So (5-7) describes a much different orthogonality between B and E than the equation (5-4), where we sourced it from. The (5-7) may provide significant insights in better conceiving how waves propagate, as follows:

A comparison between (5-6) and (5-7) reveals that a swapping has taken place between the parameters of space (dx) and time (dt) in the denominators. This comparison may in fact acquire a complementary reading by swapping the denominators of (5-7) to get the mathematically equivalent expression (5-8).

$$(1/\mu_o)\, dB \,/\, (\text{-}1/dt) = (\varepsilon_o)\, dE \,/\, (\text{-}1/dx) \qquad\qquad (5\text{-}8)$$

$$dB\,/\,dt \quad = \quad dE\,/\,dx \qquad\qquad (5\text{-}9)$$

Now the left-side accounts for a change in B with respect to **imaginary time** $idt = (\text{-}1/dt)$, and the right-side accounts for a change in E with respect to **imaginary space** $idx = (\text{-}1/dx)$, where the two minus signs cancel out. Notice how (5-8) compares with (5-9) -which corresponds to (5-6) written in the opposite direction- whose one side accounts for a change in B with respect to **real time** (dt), and the other side accounts for a change in E with respect to **real space** (dx). The denominators of (5-8) and (5-9) reveal that **these two relations approximate the same type of interaction, from the point of view of opposite parameters** (the one imaginary, the other real). This assumption may also extrapolate to imply that **Maxwell's law (5-4) is an opposite energy variant of Lenz's law (5-5)**. Let us keep that in mind, as it shall be useful in conceiving effects of reaction that will be considered in section 6.

In relevance to the above, a comparison between (5-8) with (5-9) additionally reveals that when the denominators concern negative reciprocals of space and time, that necessitates the involvement of the electric permittivity (ε_o) and the magnetic permeability (μ_o) for maintaining equivalence. Or, in different words, to measure a change in B with respect to imaginary time -$(1/dt)$ instead of real time (dt) requires the involvement of $(1/\mu_o)$, and, to measure a change in E with respect to imaginary space -$(1/dx)$ instead of real space (dx) requires the involvement of ε_o. Therefore, in a way, the involvement of ε_o and μ_o acts as a bridge between the negative energy and the positive energy domains. We shall soon figure why this is so. It further appears, that the role of ε_o and μ_o is not that of a simple "adjustment factor". It is instead far more fundamental and subtler, as ε_o comes in units of Farad per meter, and μ_o comes in units of Henries per meter. This tells that the change of the electric field *per imaginary space* $[\varepsilon_o dE/(1/dx)]$ implicates capacitance C (contained in ε_o), while the change of the magnetic field *per imaginary time* $[(1/\mu_o)dB/(1/dx)]$ implicates inductance L (contained in μ_o). Furthermore, both C and L are known to be complex entities (each having a real as well as an imaginary coordinate) [13], and that means that they may engage in distinct, orthogonal interactions.

It appears that the deeper nature of C and L had been poorly understood so far. The theory of electronics associates them to a phase lead or lag between the oscillations of voltage and current. In the theory of electronics, the "lag" is attributed to the time needed for a magnetic field to change, but no similar explanation may apply for the "lead" in time, leaving room for a differentiated reasoning behind the essence of both shifts. As per the theory provided here (including the next sections), the capacitive lead corresponds to a non-classical phase shift forward in time $(+\Delta t)$, while the inductive lag corresponds to a corresponding phase shift backward in time $(-\Delta t)$. In relation to that, the electric permittivity and the magnetic permeability may be expressed as indicated in (5-10) and (5-11):

$$\text{Electric permittivity} \qquad \varepsilon_o = (C/dx) = +\Delta t/dx \qquad\qquad (5\text{-}10)$$

$$\text{Magnetic permeability} \quad \mu_o = (L/dx) = -\Delta t/dx \qquad\qquad (5\text{-}11)$$

A point of attention is that a "shift *in* time Δt" could be measured as an angle of phase shift, but could alternatively be expressed in seconds, in which case the units look similar to the units used for a change "*with respect to* time" dt, which is also expressed in seconds. We shall refer to this point often, as we move ahead. The overall supposition that the essence of C and L refers to phase shifts in the time domain (taking place with respect to space), could be put in test on the expression (5-12) which specifies the speed of light. If the supposition is correct, the units of the right side of the equation should match the ones of velocity at the left side of it, referring to dx/dt.

$$c = (\varepsilon_o\, \mu_o)^{-\frac{1}{2}} \qquad\qquad (5\text{-}12)$$

Here, the units in the parenthesis $(\varepsilon_o\, \mu_o)$ refer to $((+\Delta t/dx)(-\Delta t/dx)) = \Delta t^2/dx^2$ thus the right side of this relation yields $dx/\Delta t$, which matches the units of velocity dx/dt at the left side, both measured in meters/sec, subject to the inaccuracy that Δt is a "phase shift in time" so it has different essence to dt. That the units are same in both cases, allows to (loosely) pass this check, however, this inaccuracy brings into surface a first hint, that the propagation of light may involve a certain process taking place in the time domain.

Notice also, that the particular units used for ε_o and μ_o, both measured in sec/meters, allow to explain why the denominators of (5-8) necessitate the involvement of the electric permittivity (ε_o) and the magnetic permeability (μ_o). While (5-8) describes field alternations with respect to *imaginary* time and *imaginary* space, the involvement of ε_o and μ_o gets the units in each side of (5-8) turn to resemble units of an equation describing positive energy. That is, the units in each side become comparable to the units of (5-9). In effect, the involvement of ε_o and $1/\mu_o$ gets equation (5-8) reduce to resemble (5-9), subject to the inaccuracy that Δt and dt share same units without being truly identical quantities (since the former refers to a shift). And inn relation to this, it further arises that equation (5-9) is unable to account for any possible course of action taking place in the time domain during the propagation of an EMW. **It describes only what we perceive in the real world of classical physics, which concerns alternations with respect to specific values of "fastness of passing time" (*dt*) and "stretchiness of space" (*dx*) which prevail in our cosmic region** (referring to point S of figure 1-1).

Equation (5-8), instead, reveals that each alternation of the electric field E goes along with a concurrent shift $+\Delta t$ in the positive direction of time (due to the co-involvement of ε_o), and each alternation of the magnetic field B goes along with a concurrent shift $-\Delta t$ in the negative direction of time (due to the co-involvement of $1/\mu_o$). The shifts are unidirectional, and this makes the wave propagate via repetitive unidirectional displacements in time (to be further elaborated, shortly). In fact, it is not the electric and magnetic fields that alternate between each other. The involvement of ε_o and $1/\mu_o$ gets the actual fields involved to concern the **displacement electric** field (5-13) and its magnetic counterpart which we previously referred to as the **displacement magnetic** field (5-14), which have fundamentally different attributes.

$$D = \varepsilon_0 E \qquad\qquad\qquad (5\text{-}13)$$

$$H = (1/\mu_0)\, B \qquad\qquad\qquad (5\text{-}14)$$

A different way to arrive at the same conclusion, is to take into consideration that what we consider to be an alternation of B as per (5-15), is actually an alternation of D as per the relation (5-16):

$$B = (1/\upsilon)\, E = (dt/dx)\, E \qquad\qquad\qquad (5\text{-}15)$$

$$D = \quad \varepsilon E \ = (+\Delta t/dx)\, E \qquad\qquad\qquad (5\text{-}16)$$

And likewise, what we consider to be an alternation of E as per (5-17), is actually an alternation of H as per (5-18), shown below. In fact, the right side of (5-18) involves a shift backwards in time (referring to the minus sign in the term $-\Delta t$), and since negative-time corresponds to an imaginary entity, the magnetic permeability projects toward the real world in the form of a negative reciprocal $-1/\mu_o$. (The second negation concerns a counterbalancing character, so the two negative signs may cancel out).

$$E = \quad \upsilon B \ = (dx/dt)\, B \qquad\qquad\qquad (5\text{-}17)$$

$$H = -(1/\mu_o)\, B \ = -(dx/-\Delta t)\, B \tag{5-18}$$

Therefore, equation (5-8) -as well as (5-7) from which it arose- is more properly expressed through the form (5-19), or rather, its equivalent (5-20).

$$dH/-(1/dt) = dD/-(1/dx) \tag{5-19}$$

$$dH/idt = dD/idx \tag{5-20}$$

At this point, it is of significance to apprehend how do the displacement fields H and D displace: According to the theory presented here, any effect of reaction has imaginary characteristics with respect to the effect of action that it reacts. Considering that an imaginary coordinate axis is mathematically orthogonal to a real coordinate axis, this translates in that, **the reaction should deploy in an orthogonal direction** with respect to the action it reacts. In relation to that, since an EMW is generated in *reaction* to charge's acceleration, the displacement of fields D and H should trasmit sideways to an acceleration of charge. In connection to this, we need to address two points of significance, one referring to the denominators of (5-20), and the other referring to its numerators:

(i). <u>Denominators</u>: Recall that earlier we made a swapping of the denominators of (5-7), which brought up (5-8) and eventually (5-20). In quantitative terms this swapping may seem like simply expressing the same equation in a different form. In physical terms, however, **the resulting equation describes a completely different condition, since it accounts for the alternation of fields with respect to imaginary time (idt) and imaginary space (idx), and this refers to an orthogonal coordinate setting**. In relation to that, the shift should deploy over the coordinate setting were gravity and mechanics apply, where the Weinberg angle θ_w develops (e.g. the shift of the time metric with respect to the space metric shown in figure 1-1). This suggests that the phase shifts in time which are accounted for through the permittivity and permability terms in electromagnetic wave propagation are not concerning the conventional inductive and capacitive shifts met in electronic circuits, where the shift develops in the direction of acceleration of charge. Now the shift develops in the orthogonal direction, therefore toward the radiative direction (at right angles to the acceleration of charge). In such case, the shift rather applies as a shift of the magnetic coordinate, over space (where the Weinberg angle develops). In that sence, **an EMW may be understood to propagate through repetitive step-by-step unidirectional displacement shifts, per field every field alternation.**

(ii). <u>Numerators</u>: Since the displacement fields D and H contain the terms ε and μ, and the latter can be complex numbers (having a real as well as an imaginary coordinate), this means that the displacement fields D and H can also be complex, each having a Real and an Imaginary coordinate. This also implies that they may engage in orthogonal interactions. In relation to further material that will be considered in section 6, the propagation of an electromagnetic wave concerns an alternating displacement on space between a cosine coordinate of the displacement electric field $D=\varepsilon E$ and a sine coordinate of the displacement magnetic field $H=(1/\mu)B$, with respect to the Weinberg angle. Note that since the magnetic field is by default orthogonal with respect to the electric field, an additional imaginary term gets involved on H turning that to Real as well, as per the relation $i^2=-1$. The corresponding alternation of D and H with respect to idx and idt complies to the representation of the photon $\gamma=\cos\theta_w\ \mathsf{B} + \sin\theta_w\ \mathsf{W}_3$ in the electroweak theory, which shall be considered in section 9. The repetitive shift in the space domain by θ_w actually allows for the magnetic coordinate of radiation to convey a form of "displacement-momentum" of negative energy, in analogy to as a shift in time allows for the electric coordinate of radiation to to convey "displacement-current" over the plates of a capacitor in electronics. And as we shall figure, this is what makes it possible for photons to be able to change the energy state of an electron, which concerns the electron's matter-related energy (not charge-related).

All the above, therefore, seem to provide new insights on EMW propagation. Conventional equations of classical physics miss to take into consideration the involvement of a shift in the time or space domains, so they describe the alternation of fields with respect to **unitary** (fixed) values of the metrics of space and time which prevail at our cosmic neighborhood, meaning specific fastness of passing of time and specific space stretchiness, used in this equations' denominators dt and dx. This is a partial representation of an EM wave's propagation, that only accounts for a classical-wave approximation to the wave's propagation. Instead, equation (5-20) actually involves the phase-shift (displacement) in time which the propagation of waves co-engages. The new context allows for a classical-physics-way to intuitively approximate the quantum-physics-related behavior of radiation, since the displacement fields interact in **non-unitary** values of time and space. That actually is responsible for bringing the quantized behavior of radiation in the form of quantized lumps (the photons), which shall be discussed in sections 7-9. As we shall figure, an involvement of a *range* of fastness of time incorporated in an electron's wave function lets elementary particles perceive EM waves like a collection of waves of shorter and longer wavelengths of diminishing strengths, which, according to Fourier analysis behave as lumps of energy, the photons.

On a point of interest, we have just described that the propagation of an EMW concerns an alternation between a cosine coordinate referring to the field D and a sine coordinate referring to H. In that case, it is of reason to also consider what holds for the other coordinates, meaning the sine of D and the cosine of H. That appears to be tricky:

- We have pointed out a difference between the inductive/capacitive behavior in an electric circuit, and the inductive/capacitive behavior accounted for through the permittivity and permeability terms in EMW propagation. The former describes what holds along the direction of acceleration of charges, for example during the transfer of displacement current as per $I_d=\varepsilon_o d\Phi_E/dt$ between the plates of the capacitor, which is what the right side of (5-4) describes. It should concern a negative energy (imaginary) variant of the electric field, it should not be radiative, and should damp out quickly. Instead, (5-4) seems to involve only Real terms. It turns out that these terms appear to be real due to the -already exisiting- negative reciprocation of the magnetic-field-related term $(1/\mu_o)$ and concurrently due to the -already exisiting- reciprocation of the parameter of space in the electric flux term Φ_E (since area goes to the numerator of the flux term). Such reciprocation gets the relation (5-4) appear like being real, while in fact it describes a negative energy effect. Hence the corresponding fields become evanescent at close range (thus the plates of a capacitor need to be in close proximity for the interaction to be in effect).

- While the displacement current through the plates of a capacitor involve a negative energy (imaginary) variant of the electric field, the transmission of the EMW appears to involve a negative energy (imaginary) variant of the magnetic field. In that case, the coordinate setting with respect to idt and idx appears to allow for displacement magnetic coordinate to convey an analogy of "displacement momentum" p_d instead. In section 9 it shall actually be shown that the Z^0 boson also involves the transfer of reactive **displacement-momentum**, providing a new interpretation of how **the Z^0 boson illusively appears like carrying mass**, while it actually mediates a reactive process which transfers a negative form of momentum, instead. The negative energy of such particle's fields makes it damp out really quickly in the classical world, just like evanescent wave modes also do, this is why the Z^0 particle has a very short range of interaction. Photons instead, will be discussed to carry **"negative" displacement-momentum** where that is an imaginary entity. Because of that, photons appear as being massless with respect to our real world. While with respect to the imaginary portion of an electron's wave function, they may transfer negative displacement-momentum at long range, and charge the (matter-related) energy state of an electron.

<u>**Point 3**</u>: **Magnetic force action and reaction**

We shall now address how the magnetic field applies its force as per the relation $F=q(\upsilon{\times}B)$, which is something that the Maxwell equations may not address. Along with that, we shall be looking for answers to questions like:
- Why the magnetic field does not seem to get "reduced" in some way, when it has affected a magnetic force.
- How is an opposite force transferred on a magnet's atoms or molecules, when that magnet exerts a magnetic force on a moving charge, for momentum to be conserved.
- How is the magnetic field involved in the creation of an inductive phase lag (between voltage–current oscillations).

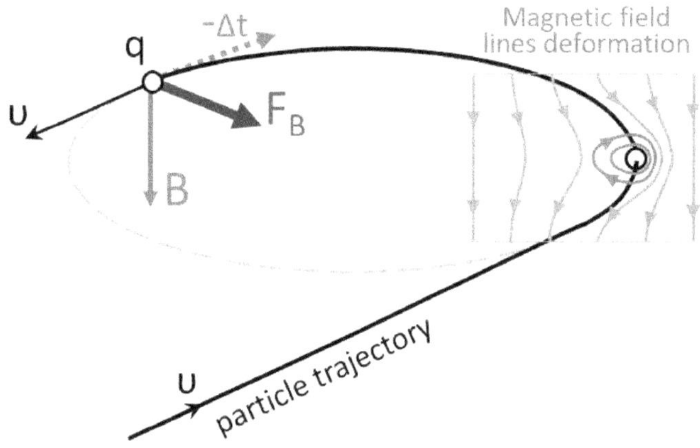

Figure 5-6: Cyclic trajectory of a moving charge,
once it gets under the influence of an external magnetic field

We shall seek answers to these questions via reference to figure 5-6, which depicts how a magnetic force is exerted upon a charge that moves within an external magnetic field. The magnetic field lines in this figure are supposedly oriented from the top to the bottom of the figure. In a preliminary level of understanding, the charge is considered to be deflected in relation to a **sideways deformation** (sideways shifting) of the field lines of both the external magnetic field, as well as of the magnetic field of the moving charge itself (see window/insert at the right side of figure 5-6). In this case, the field lines of both these two magnetic fields shift toward opposite directions, therefore exerting a displacement force on the moving charge and at the same time an **opposite** force to the external magnetic field.

A point of attention is that the magnetic deflection force changes the direction of motion of the charge WITHOUT changing its kinetic energy. In this sense, the magnetic force applies much like a centripetal force acts on a rope which keeps an object in circular motion in mechanics. Note also, that the magnetic field gets the charge to follow a circular path, which is a same form of path that a charge would follow if it was supposedly moving through the turns of it coil (of same radius). In either case, the sideways deflection of charge requires an opposite counterbalancing effect. Current theory associates the cyclic motion of charge moving through a coil with a value of inductance of the coil. However, there is currently no deep intuitive clue about how this association materializes and how a phase lag develops, other than a (rather poor) explanation suggesting that the lag concerns the time needed for the alternating magnetic field to change. This explanation however seems inadequate, and we shall provide a different approach on what seems to actually hold, where that will concurrently address pretty well the questions raised above. We shall assume that during the action of a magnetic force two different processes take place at the same time, which are in balance to each other and either one is imaginary with respect to the other. We may first outline these two processes as follows, and shall consider them at larger depth right next:

- The first of the two co-involved effects concerns a *real action*, the actual outcome of the magnetic force, which may deliver physical result (displacement of the charged particle) at right angles to the magnetic field lines. The strength of the force associates to the difference in the **density** of the magnetic field "lines", which act as forcing the charge in the direction from the denser toward the less dense lines. That is, the magnetic force appears to arise from a shift of the magnetic field lines, where that, corresponds to a shift in negative reciprocal (imaginary) time, in the *sideways* direction (and not along the direction of the magnetic field arrows, where that would constitute a change in FLUX). Since this "density-varying" vector intersects nearly orthogonally to the lines themselves, this gets the magnetic action to take place without consuming the energy of the magnetic field itself (without affecting its flux) in the first place. Instead, the magnetic force utilizes energy in an orthogonal direction (orthogonal coordinate) as we shall describe next. Furthermore, as the magnetic deflection force gets the moving charge to follow a circular path (just like a rope would keep an object in circular motion in mechanics), we may assume that the magnetic force's action has a RADIAL-like character (centripetal-like). In this respect, due to that radial-like geometry of the interaction, we may consider that the magnetic action resembles a **capacitive-like** process.

- The second co-involved effect, concerns an *imaginary reaction*. This gets to have an energy-counterbalancing role which is in need for conservation purposes (since the magnetic field itself does not get "consumed"). As per the reactive nature, and corresponding imaginary character of any effect of reaction, **a reaction should deploy at right angles to the effect of action that it reacts**, just like an imaginary axis deploys at right angles to a real axis. In fact, the reaction develops at right angles to both the direction of the magnetic field lines B (which correspond to equipotential lines, and stay as such) and to the direction of the magnetic force F_B. So, in this case the direction where the reaction is to take place identifies to the general direction of the motion of charge (or electric current if the charge belongs to a current). But since the sideways deflection of charge does not change the particle's kinetic energy, the reaction process needs to borrow energy in some other way. This is done by borrowing energy in the time domain. This concerns an inductive type of effect, which generates the inductive phase lag (vector $-\Delta t$). **The corresponding borrowing of energy in the time domain is what brings the lag of current with respect to voltage in the case of an electric circuit oscillation**.

In such way, the effects of action and reaction develop in opposite sign of energies, so that they get to counterbalance each other. They also get to materialize in orthogonal physical directions, just like an imaginary coordinate axis is orthogonal to a real coordinate axis. Let us now consider these two parallel processes in certain more detail:

A. The Action part of the magnetic interaction.
The exertion of a magnetic force on a moving charge is known to arise as the motion of the charge deforms the density of the magnetic field around it, and as a result the moving particle is being forced in the direction from where the magnetic field "lines" are denser, toward the side where they are less dense. The corresponding force is described by the relation $F=q(\upsilon \times B)$, where q is the charge, υ its velocity, and B the external magnetic field. A visual comparison of this equation with that of the electric force $F=qE$ provides a hint that in the magnetic case the "force-field agent" is the combined term $(\upsilon \times B)$, and not the magnetic field B on its own. This may also suggest that that what we call a magnetic force is not a force of B, it is rather a force of $(\upsilon \times B)$. At this point we need to recall what was mentioned a little earlier, that the units of velocity $\upsilon=dx/dt$ resemble the units of the reciprocal of the magnetic permeability $(1/\mu_o)=dx/-\Delta t$. In such case, we shall focus on the point that the motion of

charge through an external magnetic field creates a displacement (a sideways shift) of magnetic field "lines". This shift of the field lines involves units which resemble the units of velocity. The shifting of the magnetic field lines corresponds to a shift in the time domain, and since the magnetic field applies as a reaction type of field, this should concern a shift in the *negative* direction of time, thus an inductive shift, denoted as $L = -\Delta t$. Since that shift is developing with respect to space dx, it may actually be represented in the form of $-\Delta t/dx$ where that matches the units of the permeability term $\mu_0 = -\Delta t/dx$. As the generation of the sideways shift of the magnetic field "lines" is an effect of reaction, it should concern negative energy, it should therefore have imaginary character. And as aforementioned, an imaginary entity projects to the real world through its negative reciprocal, since $i^2 = -1$ so $i = -(1/i)$. Therefore, in such case the permeability term should project in the real world in the form of negative reciprocal $(-1/\mu_0)$ instead of the original μ_0. And indeed it does get involved in reciprocal form in corresponding formulas. This suggests that the motion of charge generates a term $(-1/\mu_0)$. In such case, the difference between the reaction born $-1/\mu_0$ (which corresponds to units of $dx/\Delta t$) and the charge's velocity which *caused* it (where the units of velocity correspond to dx/dt), reduce to a difference between the parameters Δt and dt, both of which may be measured in units of seconds. In relation to that, what we have been considering as a magnetic force $F = q(\upsilon \mathrm{x} B)$ seems to correspond to a force of the form $F = q\ ((-1/\mu_0)B)$. And since the term $((-1/\mu_0)B)$ identifies to the displacement magnetic field H, the equation describing the magnetic action force could be approximated as follows:

$$F = qH \qquad \text{[approximation]} \qquad\qquad (5\text{-}21)$$

While the above concept may seem to make "some" sense, and might seem okay in terms of units, it is not exactly accurate though. The reason is, that H includes the magnetic permeability μ, and that in turn includes inductance L, which corresponds to a negative shift in time $-\Delta t$. But such a shift typically develops loop-like (for example, it identifies to the phase lag over a circuit loop). While in the case of magnetic action we have a differential in time which develops RADIAL-like (mimicking the force on the rope in the mechanical case or an object that circulates), so the magnetic action seems to have a *capacitive* character (instead of an inductive one).

That calls for addressing TWO types of displacement of magnetic field, both of which have similar raw units, but they come in opposite signs in terms of displacement. We shall denote them by the letters $H-$ and $H+$, where the negative sign $(-)$ refers to an "Inductive displacement Magnetic field", and the positive sign $(+)$ refers to a "Capacitive displacement Magnetic field".

$$H- = (-1/\mu_o)B \qquad\qquad (5\text{-}22)$$

$$H+ = (-1/\varepsilon_o)B \qquad\qquad (5\text{-}23)$$

The μ_o corresponds to $-\Delta t/dx$ while ε_o corresponds to $+\Delta t/dx$, so they are seemingly similar entities that differ in the sign of the phase shift. Their opposite sign also demands that they operate in opposite energies, which implies that either one of the two should operate in imaginary conditions with respect to the other. That in turn, signifies that only one of them, either μ_o or ε_o should come in reciprocal form. That is, since the permeability comes in reciprocal form $(-1/\mu_o)$, the permittivity should appear in the form (ε_o). It appears, however, that in the case of a charge's deflection, the sideways shift in the magnetic field lines is an effect that takes place with respect to the opposite energy domain. More will be said on this point in the next section, where it will be shown that (5-22) appears to concern an interaction with the real part of the particle's wave nature, while (5-23) appears to concern an interaction with the imaginary part of the particle's wave nature. Hence, this interaction should involve the permittivity in reciprocal form too, as the term (5-23) describes. In other words, since (5-22) and (5-23) are involved in interactions in opposite energy domains, they should both have the permittivity and/or the permeability terms appear in reciprocal form. While a charge is moving with **conventional velocity $v=dx/dt$, it creates an opposite potential $+\Delta t/dx$ as viewed with respect to the negative energy domain.** And that potential causes the displacement of the moving charge, without changing its kinetic energy. Consequently, the relation (5-21) may be re-written more accurately as follows, where the notation of the displacement magnetic field includes the positive mark to signify the capacitive version of displacement.

$$F = q\, H+ \qquad\qquad (5\text{-}24)$$

At this point, it is of interest to highlight the relative resemblance between the displacement magnetic field $H+$ and the electric field E, as it was also described in the comparison between (5-17) and (5-18) of above. That is, $H+$ and E resemble each other in a pretty much similar sense as (vxB) resembles (vB). And **the force that is currently called magnetic, actually concerns a "capacitive displacement magnetic" force.** The way that it acts, concerns a capacitive shift (a shift in the positive direction of time), as this is viewed with respect to the imaginary portion of the charged particle's wave nature (to be addressed in section 8). It is therefore a sort of *"negative-capacitive-like displacement of the magnetic field* that is responsible for the so-called magnetic action (and not the plain magnetic field itself). This gets the action of $H+$ to resemble an electric field's action, forcing the moving electron sideways, as shown in figure 5-7, left part (force arrows). Notice how the field lines of the field $H+$ (at the right part of figure 5-7) look similar to the field lines of an electric field between two charges of opposite sign. In particular, the one pole of the $H+$ reflects the concave bending of magnetic field of the moving charge (white dot), and the other pole of the field $H+$ reflects the convex bending of the magnetic field of the external magnet (hypothetical black dot).

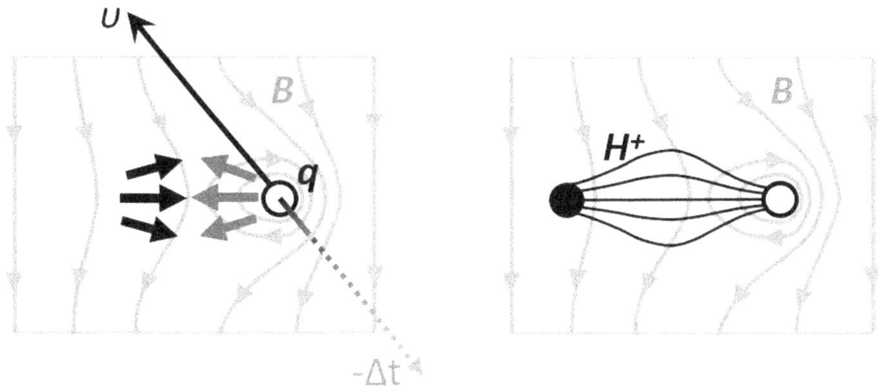

Figure 5-7: *Left*: Deformation of magnetic field lines B by a moving charge, creating a capacitive-displacement-magnetic potential (bold arrows).
Right: The capacitive displacement magnetic field lines $H+$ resemble the field lines of an electric field between two particles of opposite charge

According to the above, **the field lines of $H+$ correspond to trails of phase lead In the time metric, as that is viewed from the imaginary point of view, so it resembles trails of curvature in negative reciprocal time. Furthermore, as these have a capacitive character, the field lines of the conventional magnetic field B serve as equipotential lines (or equipotential surfaces) of $H+$.** That the field lines of B only represent equipotential surfaces, and the magnetic force develops almost orthogonally to them, is actually the reason why the magnetic field B does not get "consumed" when a magnetic force acts. While the magnetic force acts, these surfaces are shifted, but they still represent equipotential surfaces. Actually, this is why the difference in the density of magnetic field lines (being denser in the one side of the charge and less dense from the other side of it) relates to the direction and strength of the magnetic force. And while the sideways shift of the magnetic field concerns a phase shift in imaginary time, the charge's deflection (a sideways centripetal-like acceleration) gets to release the corresponding potential through a similar Doppler-like process, in analogy to as mentioned earlier in section 5.1 to hold for the gravitational acceleration. In other words, **while the charge becomes deflected, the potential associated to the phase shift (in imaginary time) of the displacement magnetic field, and a Doppler-like differential (in imaginary time) due to the charge's centripetal-like acceleration, balance out each other. This releases energy, and this release is what gets a moving charge to deflect** all by itself (for the same simple reason to as a compressed spring would stretch out all by itself to release its compression).

Notice that one key difference between a electric force (driven from a curvature in negative reciprocal space) and a displacement-magnetic force (driven from an effective shift in time as that is viewed from the imaginary domain) is that, in the former the curvature in negative reciprocal space is permanent (concerns the force by a static electric field of a charged particle or object), while in the later the phase shift in imaginary time develops only if the charge is moving. In relation to that, the electric force may accelerate a charge and change its kinetic energy, while the displacement character of the magnetic force generates acceleration sideways to the direction of motion of charge, so it only deflects its direction of motion without changing its kinetic energy.

As we move on, we shall actually figure that it similarly works for other fields and forces in general, including mechanical forces like the Bernoulli pressure which we shall consider in section 5.4. That is, **all forces concern either curvatures in metrics (like the curvature of time, or of negative reciprocal space) or phase shifts (leads or lags in time or space) which cause deflection of particles**. The curvatures in metrics (in "our" cosmic region) typically apply in positive energy and should be able to change particles' kinetic energy, while the phase shifts typically apply in opposite energy and should in principle deflect particles without changing their kinetic energy.

B. The Reaction part of the magnetic interaction.

As mentioned above, the magnetic deflection of a moving charge (illustrated in figure 5-6) and the associated sideways acceleration, is symmetric to the exertion of centripetal acceleration in mechanics. In this respect, the deflection of the moving charge may be perceived as a "load", which the magnetic force gets to displace sideways. And while charge does not exhibit mechanical inertia when it is deflected, there seems to be a similar impeding process involved during its sideways acceleration. And that impeding process appears to yield a dragging-like reaction. Let us explore that step-by-step:

Recall that the magnetic deflection arises when there is an asymmetric change in the density of the magnetic field lines (for instance, the lines becoming denser in the one side of the moving charge and less dense in the other side of it). The creation of this asymmetry in the density of lines requires energy, so that energy should be sourced in some way. Considering that the asymmetry was caused by the motion of the charge, the sourcing of the energy should rather come from the source that powers the motion of the charge. Supposedly that the moving charge is part of an electric current, then the source of its motion is the circuit's electric battery. In such case the reaction should be transferred to the battery, and that should be done through a process that allows for borrowing energy from the battery in order to be able to cause the asymmetry in the density of the magnetic field.

As the asymmetry in the density of the magnetic field lines was stated to concern a shift in negative reciprocal time, the reaction should rather be conveyed to the battery through an effect in the time domain. The way that this takes place is by affecting a type of "time-dilatedness" to the moving charge (or electric current), where that corresponds to the phase shift back in time $-\Delta t$. This is what we perceive as an inductive phase lag in electronic circuits, which gets an oscillation in current to lag an oscillation of voltage. Notice that such kind of shift concerns a "displacement", not a "jump" in time. Such inductive shift in time gets a portion (an imaginary coordinate) of the battery's supplied energy to not do work, due to its imaginary nature, so it applies as an inefficiency of the battery to provide work. However, the corresponding energy is not being lost. Through the inductive shift in time, **the kinetic energy of the moving charge stays unchanged, but the displacement BORROWS energy from the battery and provides it in the form of potential energy in displacing the charge's magnetic field, and by extend in generating the magnetic force**. So, in a loose sense, the borrowing of energy from the battery (through the shift) plays a role like that of the strain on a rope that is used to rotate an object circularly in mechanics.

Notice that (in the case of a circuit) the inductive phase shift in current is mediated throughout the whole circuit loop simultaneously. This means that the reaction is affected on the battery (and any other part of the circuit loop) immediately. **This form of instantaneity is feasible since the case does not involve conventional movement of charge, instead it involves a concurrent displacement (shift) on the whole loop (in time)** just like when turning a bicycle's wheel all the elements of the wheel rotate concurrently (instantaneously). On a side-note of significance, this feat of instantaneity may apply in an oscillatory fashion as well, making it possible to deliver instantaneous action (similarly to as this is possible in entanglement). That principle is utilized in an **invention** made by the author (not included in this book), which may provide ultrafast transmission of energy, as well as the means for ultrafast signal transmission.

Notice that in the paradigm of figure 5-6 we have two magnetic fields interacting with each other; the magnetic field of the moving charge (or electric current), and the magnetic field of the external magnet. Since both affect each other, both need to borrow energy from their source. The one borrows it from the battery, and the other borrows it from the atoms or molecules of the external magnet.

Later we shall figure that the *magnetic flux* is also a property that tends to be conserved. However, in the case of exertion of a magnetic force, the action takes place almost at right angles to the magnetic field lines, and in that sense the force should not affect the magnetic flux directly. However, the moving charge (which, for reasons of simplicity may be considered to concern part of an electric current) conveys a phase shift, and through that a reaction potential, toward the power source (the electric battery), affecting its inner process which shall be considered ahead. Concurrently, a similar inductive shift should affect on atoms of the external magnet which generate the external magnetic field.

If we were to put the magnetic force's interaction in the form of an equation, the one side of the equation could describe the change in the density of the magnetic field lines (a potential created as a result of the motion of charge inside an external magnetic field). The other side of the equation could describe the deflection of the charge, in the form of a centripetal-like acceleration (a sideways "flow") of charge, which applies radial-like. A corresponding equation that describes the magnetic force should look much like Faraday's fourth law in reverse order. This reverse order also implies an orthogonality. The shift in magnetic field is now radial-like (sideways to the direction of the magnetic field lines), and the deflection concerns motion of actual charge (not a "displacement current"). More shall be said on that ahead.

Point 4: Raw units of the magnetic field

That the magnetic field B may *not* act a force on a charge if the charge isn't in motion, may also be traced in the units of the magnetic field, which do not seem to exhibit an "acceleration-like" term, where that would imply a force action. (That's in contrast to what holds with the units of the gravitational field which have the acceleration-like form dx/dt^2, or, the units of the electric field which were explained to concern a similar "acceleration-like" term $dt/d(-x)^2$).

The units of the magnetic field are named after the inventor Tesla. But in order to obtain a deeper insight on the field's essentials we shall look for "raw units", meaning units that would rely upon raw parameters like those of space and time. To work on that, we may start with a reference to the known relation $E=vB$. (Actually, as mentioned previously the units of E resemble the units of H which could be more proper to use in representing the case, but due to the similarity we may stay with the units of E which are already familiar). In that case, the "raw" units of B should resemble the units of $(1/v)E$, which refer to $[(dt/dx)(dt/d(-x)^2)]$, and thus to $(sec/m)(sec/m^2)$, which makes for **sec^2/m^3**.

This type of units may seem bizarre and unrealistic, but we shall seek justification that this is how they are. Such a justification would come if these raw units could reach certain match to the "conventional" units of Tesla, where $1 Tesla = 1 Volt \cdot sec/m^2$. What we need to explore in this case is the units of a Volt. But this is easy to approximate, since by now we already have knowledge of the raw units for the electric field. The electric field relates to voltage (e.g. the voltage between a capacitor's plates) as per the relation $\Delta V=Ed$ where d is the separation between the capacitor's plates. According to that, the units of voltage should be corresponded to $(dt/d(-x)^2)(dx)=dt/dx$, or **sec/m**. Notice that the units dt/dx look like the conventional velocity dx/dt subject to swapping between the parameters of time and space. This is not by coincidence, as the same units resemble the ones of the electric permittivity, which involves capacitance, and is measured in terms of a phase shift in time, with respect to space. In this respect these units of Volts do seem to make sense, and that should be kept in mind for later use as well. By substituting the Volt with sec/m, we get the raw units of Tesla to correspond to $(sec/m)(sec/m^2)$, which makes for (sec^2/m^3), which matches the particular raw units of B.

Actually, when we speak of units, attention needs to be put in distinguishing between a phase shift "IN a parameter" and a change "WITH RESPECT to this parameter", for example between Δt and dt, or between Δx and dx, or their negative reciprocals correspondingly. This is so, because, even though the phase shifts (like Δt or Δx) are conventionally measured in degrees of angle, they may also be approximated in terms of seconds or meters. For instance, the electric permittivity $\varepsilon_o=+\Delta t/dx$ and the magnetic permeability $\mu_o=-\Delta t/dx$ could be measured in seconds per meter (aside the sign of the shift's direction). This also applies for the units of displacement fields, like for instance the units of the displacement magnetic field H. The latter was explained to come in two versions, a capacitive one for which it holds $H+ = (1/\varepsilon_o)B$, and an inductive one where $H- = (1/\mu_o)B$. In such case, the units for both versions of H should be $(m/sec)(sec^2/m^3)$, which makes for **sec/m^2**. Notice how this resembles the units of the electric field, which justifies that H may apply as a radial-like field, and the magnetic force that it produces resembles an electric one as explained above (and as the symmetry between the corresponding relations $F=qH+$ and $F=qE$ indicates).

An additional point, which is of significance and may serve for later use, is that the units of B may also be re-written differently, suggesting an engagement of this field in an additional way of interaction. This is done by splitting the original term (sec^2/m^3) to $(sec^2/m)(1/m^2)$, which may also be expressed in the alternative form $(dt^2/dx)(1/dx^2)$. The later includes two terms; an acceleration-like term (dt^2/dx) and a flux-like term $(1/dx^2)$. This reading may reflect a different symmetry, which additionally reveals an effective complementarity between the electromagnetic and gravitational interactions. This is so, due to the fact that the acceleration-like part (dt^2/dx) resembles a **negative reciprocation of the units of the gravitational** field, since $d(1/x)/d(1/t)^2=(dt^2/dx)$, referring to an imaginary version of that field. At the same time, the other term $(1/dx^2)$ describes a **flux** (actually a flow, since dx^2 is in the denominator position) with respect to positive space. This seems to indicate an inverse type of capability of the magnetic field, since the resemblance with imaginary g lets account for **the transfer (the flux) of negative momentum**. In a sense, the magnetic field seems to be taking the role of "imaginary gravitational flux", which could be represented as $B = \Phi g_{im} = g_{im}(1/dA) = \bar{g}(1/dA)$, where A corresponds to a surface through which the $\bar{g}$ field flows. This is a significant feat, since (in relation to material to follow) this capability of the magnetic field is what allows for a photon to change the energy state of an electron (which is matter-related, not charge-related), as if the photon carried "pseudo-mass". Another embodiment of such feat will be presented later to concern the electroweak interaction, in the case of the Z^0 boson which seems to illusively appear possessing mass, while it actually carries what we may call an "effective mass", or pseudo-mass.

Point 5: Classical vs. Intrinsic magnetic field

As per the illustration of figure 5-5, even though electrons may appear as electric monopoles (of negative charge), they also have an imaginary (hidden) part (of negative anticharge) which associates to imaginary attributes of the particle. As described earlier, the conventional real pole of negative charge associates to the electric field of the particle, which has radial-like field lines, of which we perceive only the real portion extending from the particle toward infinity, and we do not perceive the remaining part which is imaginary to us. We also considered that the imaginary (anticharge-related) part of the electron may become involved in real world's interactions through a Sine or Cosine coordinate, and in this way exert an electromotive potential (to be elaborated in section 6). This imaginary portion interacts in the real world through an inverse geometry, as it comes along with a loop-like phase shift hidden within a permittivity or permeability term. It's because of this shift that the field lines involved in the electromotive potential form loops instead of being radial-like.

It turns out that a similar type of inverse geometry may apply for the magnetic field, which also exists in two forms. The one form concerns the ordinary magnetic field, associated to the motion of ordinary charge, whose field lines form loops around a current carrying wire. The other form concerns an imaginary field version, which we may denote by B_{im} or $\bar{B}$, which becomes involved in effects of negative energy, in association to the imaginary portion of particles' nature, in a way that will be clarified ahead. That version of the magnetic field becomes involved in real world's interactions through a Sine coordinate, or a Cosine coordinate in an inverse case, therefore affecting partially an interaction. Due to the inverse geometry that imaginary properties require, its field lines are **radial**-like, representing the field of a "**magneto-motive**" potential (in contrast to the loop-like field lines of the conventional magnetic field). In this case, we are able to perceive only the portion of the magneto-motive potential's field lines which project upon the real world, and we do not perceive the part which is imaginary to us (this is somewhat similar to the case of the electric field of the electron). As per the above, the magnetic field may come in either of two complementary ways:

- One way is when the magnetic field is generated from the actual physical motion of charged particles like the electrons, as per their "particle nature". In this case the created magnetic field lines form loops around the flow of current.

- The other way is when the magnetic field is born through the wave nature of quantum particles. In association to material that shall be discussed in section 8, this involves an "effective" motion of charge within elementary particles, where such effective motion will be described to concern an oscillatory behavior of the space and time metrics which the particle's wave function seems to incorporate. In such case, magnetic properties of the particle may interact in the real world partially, through a Sine coordinate. Such sine applies with respect to a phase angle θ_w where that does not concern a lead or lag as those developed in an electric circuit. Instead, it corresponds to the 30° phase shift between the space and time curvatures depicted in figure 1-1 (the Weinberg angle). The projection of the corresponding imaginary coordinate to the real world comes through the factor of $Sin(30°)=\frac{1}{2}$. It is this value that gets a particle's magnetic field behave as having a **magnetic moment of ½**, and becomes involved in interactions at the atomic level in the form of the so-called **intrinsic** magnetic field. Due to its imaginary character, it comes with inverse geometry, with this version of the magnetic field having radial-like field lines (instead of the conventional loop-like lines of the magnetic field), and applies as a **magneto-motive** field. In fact, the radial-like character of this field version makes it look much like a field of a magnetic monopole, with another pole being imaginary and hidden behind it. This specific field version also constitutes a reason why the magnetic field of a **bar magnet** (which is mainly intrinsic-spin related) appears like coming out of a North pole and moving in a South pole, with both poles having radial-like field "lines", and the portion of the field lines from the South to the North pole supposedly being imaginary and hidden from perception. That is, ss if the one pole is a North monopole (with a hidden anti-North pole behind it) and the other is a South monopole (with a hidden anti-South pole behind it) and the portion of the field lines between the hidden anti-North and hidden anti-South being imaginary and not directly perceptible. All that is unlike what holds for the magnetic field of a coil, which has loop-like magnetic field lines, which are fully real, while in that case, the coil's magnetic poles (a North and a South) are virtual (away from the coil).

At this point, it is of significance to examine why magnets always form magnetic **dipoles** (considering that no magnetic *monopole* has ever been observed), while instead, the electric and the gravitational fields may form monopoles. To approximate this issue, let us first focus on the case of the gravitational monopole. That concerns an elementary particle (or larger object) of certain mass, which exhibits a radial-like gravitational field. We shall seek to question if that is really a monopole, or if we may be looking to a small portion of a larger picture.

In relation to what was described earlier for figure 5-2 right part, the gravitational field lines of an electron spread radially outward, however the difference in the fastness of passing time that they describe changes slope after some point. So, the further away toward the distant universe, the slower the time passes, until it reaches a halt at the nodal point d' of figure 1-1. That nodal point may refer to horizons that surrounds local celestial formations made of antimatter. But at the same time, such distant places refer to an "other pole" (since it corresponds to a node). And, with a supposed large number of such celestial formations made of antimatter existing throughout the distant universe, one could assume that the "other end" of each (radial-like) gravitational field "line" of an elementary particle on earth, would eventually dive into one (out of numerous) such distant celestial objects. Under such notion, a particle like an electron does NOT exactly constitute a gravitational monopole, it rather corresponds the one pole of a more complex set up which includes opposite poles: A real gravitational pole upon the particle, and a real gravitational pole of antimatter spread out to particles at the distant universe. Figure 5-8 left part shows how the gravitational field lines of a particle (m) may have their other end "land" to an antimatter pole ($\overline{m}$) somewhere at the distant universe. And there should be a large number of such antimatter poles spread out and away over the distant universe whereupon field lines "land", corresponding to a "de-singularity" (the opposite of "singularity"), referring to the "nodal surface 2" of figure 5.3 right part. In that sense, **a conventional matter particle does not exactly constitute a gravitational monopole as it would not literally exist solely by its own**, without its field lines reaching at other poles of antimatter.

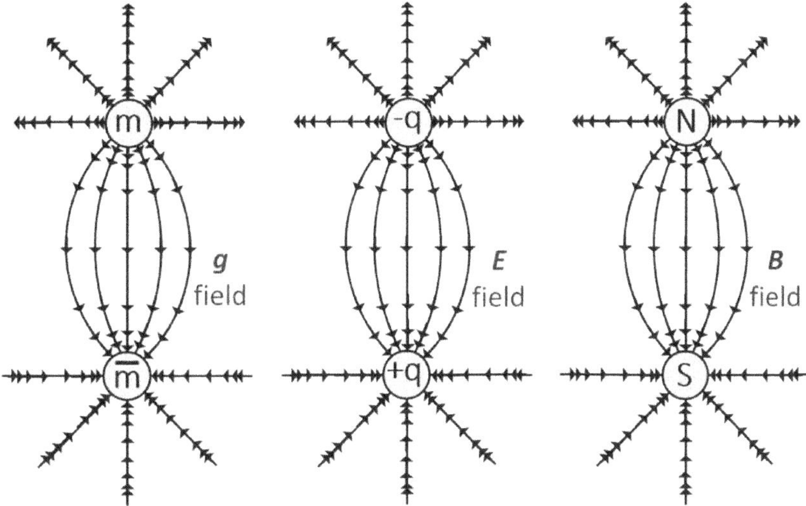

Figure 5-8: Schematic symmetry of the visible part of fields, concerning a matter-antimatter gravitational dipole (left), a negative-positive charge electric dipole (middle), and a North-South magnetic dipole (right)

On a point of clarification, figure 5-8 left part involves an antimatter ($\overline{m}$) pole, and that is fundamentally different to a "negative matter" (-m) pole. Recall that negative matter should be repulsive to conventional matter, and that should apply at close range (as close as an *imaginary* gravitational field may affect), since these two forms of matter may not coexist in particle form at the same place, for same reason that neutrinos may not reach particle form in the classical universe where electrons reach stable form. The gravitational field lines between these repelled forms of matter should resemble the electric field lines between charges of same sign of charge which repel each other.

In the other hand, matter and antimatter (which both constitute matter of positive "sign") attract each other, as per the shape of the field lines in the left part of figure 5-8. This applies similar to the attraction between a negative charge and positive charge (see figure 5-8, middle part). In such case, the radial-like electric field lines of any electron have their "other end" eventually reach into positive charge (either in the protons of nuclei, or in positrons lying elsewhere in space, as in celestial formations made of antimatter). In that sense, even an isolated electron should not be considered a monopole in the strictest sense. This is because such negative charge would not exist in nature without existence of positive sign of charge so that the two counterbalance each other. Therefore, a monopole of charge is actually one pole of a dipole whose "other pole" is spread out and away from the particle. And due to the orthogonality between gravity and electrism (the former relating to the curvature of time and the latter to the curvature of imaginary space), negative charge attracts positive charge (not anticharge), while matter attracts antimatter (not negative matter), as shown in figure 4-1.

Let us now consider if a similar condition applies with respect to a bar magnet's poles, the North and the South pole. One profound difference is that in the case of the bar magnet *both* poles look point-like, in the sense that neither is spread over the distant outer space as in the gravitational case. In addition, the two magnetic poles need to stay adjacent to each other, and it is actually impossible to isolate only one of them, since if one cuts a physical magnet (no-matter how small) in two in order to split a North pole form a South pole, each separated part will instantaneously become a new magnet having its own North and South poles. Such behavior, would require either of two conditions to hold:

(a) That each charge that participates in the generation of the magnetic field behaves as a tiny loop current. This would imply that even at the single charge's level, the magnetic field would look much like the magnetic field of a coil, so it would have two virtual poles, by default. A large magnet would consist of many such tiny loop currents, but no matter if we cut the magnet down to almost a single elementary charge level, that would also have two poles by default. A point of attention under this scenario, is that each magnetic field "line" constitutes a loop, and we directly perceive the *whole* actual loop (just like it holds for the magnetic field of a coil).

(b) That the magnetic field version involved is not classical, it rather refers to the imaginary field version considered above, referring to the intrinsic magnetic field, and has radial-like field lines. Actually, in such case each field line still forms a *hypothetical* loop, of which we perceive only the (radial-like) real portion, and we do not perceive the remaining portion which is imaginary. The field "lines", of course, are only conceptual notions, yet they are very helpful in describing a scientific model. Under the present model, in the case of a bar magnet we may perceive only the real portion of the field lines, pointing from the North pole to the South pole. While the remaining part of the field line, which supposedly points from the South pole back to the North is hidden from perception, not because the corresponding part of the field lines lies within the magnet, but because this portion of the field lines is un-real to classical perception.

It turns out that in an actual physical case, the condition (a) is not dominant in the sense that electrons in an atom do not match the classical model of a current in a loop. Instead, the approach (b) is dominant, and in order to approximate what holds in that case, we shall use a different schematic model, similar to the one we previously used to describe the gravitational field of matter. In particular, we shall use the perspective of a hypothetical side-view of each charge, exhibiting the real and the imaginary parts of a particle's magnetic field line, in comparison to its gravitational field's real and imaginary part (as originally introduced in figure 5-2, left part).

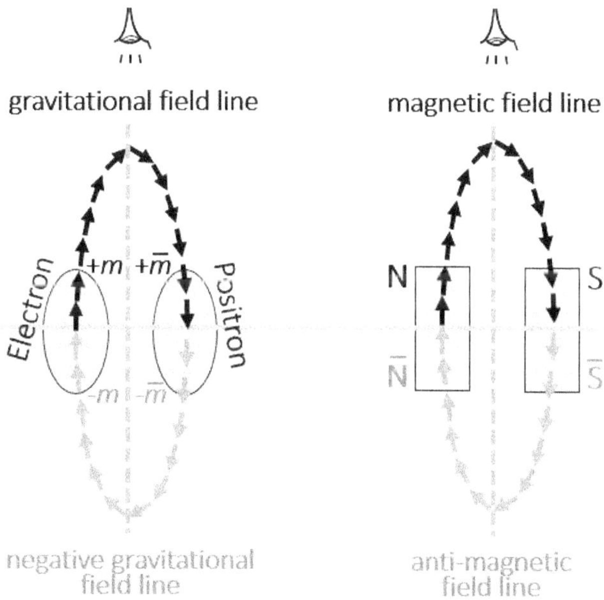

Figure 5-9: Schematic symmetry between the real field portions (black arrows at upper half) as well as the imaginary/hidden field portions (grey arrows at lower half), of a gravitational field "line" (left) and an intrinsic magnetic field "line" (right)

As explained earlier, an electron may be considered to be a hypothetical gravitational dipole, as shown in the pictorial "side-view" of figure 5-9 left part. Remember that the particles (the electron and the positron) in such schematic notion should rather be considered *point-like*, so the drawn size and shape of a particle is only for illustration purposes. (Thus, all the arrows of the field line should be considered external to the particles). According to this figure, we perceive only the visible real pole (+m) and a corresponding gravitational field line (black arrows) of the electron, and we do not directly perceive the hidden imaginary pole (-m) and a corresponding field line (grey arrows). A gravitational field line that goes radially away from an electron, may be considered to eventually reach point d′ of figure 1-1 which refers to (+$\overline{m}$) like a positron, then becomes imaginary (-$\overline{m}$) and completes a hypothetical loop to reach back at the electron. Also recall, that the part of the field line which lies to the left of the vertical dashed line ascribes to the left particle (the electron), while the part of the field line that lies to the right of the vertical dashed line ascribes to the right particle (e.g. a positron, far away).

A similar feat may hold for the poles of a bar-magnet as well, where each pole may be considered to have a hidden anti-pole of imaginary characteristics. This is illustrated in figure 5-9 right part, where the black arrows indicate the visible part of an intrinsic magnetic field "line" and the grey arrows indicate the hidden part of it. According to that, we perceive the magnet's North pole (N) and we do not perceive the hidden "anti-North pole" ($\overline{N}$), and likewise, we perceive the South pole (S) and we do not perceive the "anti-South" pole ($\overline{S}$).

If we look into the magnetic properties of an individual particle like an electron, the hypothetical side-view of a magnetic "dipole" N-$\overline{N}$ or S-$\overline{S}$ seems to reflect the so-called **intrinsic** magnetic moment, whose behavior is counterintuitive under classical physics terms. It seems that the imaginary side of the field interacts in negative energy, and projects to the real world through a sin30° coordinate (where sin30° = ½) which gives it the ½ spin factor. Its negative energy gets it interact as a **"magneto-motive"** field and potential (instead of an ordinary magnetic field), and it's an associated displacement in negative reciprocal time in the particle's wave nature that lets it interact through an inversed geometry as a radial-like field. This makes an elementary particle like the electron to behave as a magnetic monopole, a "quasi-magnet" of which we perceive only the one *real* pole (being either N or S depending on the particle's spin), with an imaginary pole ($\overline{N}$ or $\overline{S}$ respectively) hidden behind. Such real pole is not inherent (fixed) to the particle (unlike *charge* poles which have fixed sign). Instead, it is possible to flip between N or S upon change in the particle's spin. In relation to that, if a bar-magnet is cut in half, it requires less energy for half of its charges to flip than to stay unchanged, in order to minimize the total local magneto-motive potential. And that gets each elementary particle quickly adjust, by turning a cut piece of a bar-magnet into a new magnetic dipole. What this suggests, is that **the North and South poles of a bar magnet reflect the (opposite) intrinsic spin of the majority of their charges**. The same property that turns even a tiny bar-magnet into a new magnetic dipole, applies until an individual particle level. Then, since an individual electron behaves as an intrinsic-magnetic monopole, this gets individual electrons to pair up and combine opposite spins, so as to balance out the total local magneto-motive potential. It's like having each pair of electrons resemble a magnetic dipole, at the smallest level. The potential that mediates this interaction is the magneto-motive potential as stated above, and the corresponding forces refer to the so-called "exchange forces". Opposite quasi-magnet dipoles (corresponding to electrons of opposite spins) participate in **chemical bonding** interactions where they attract each other and stay together. In the contrary, pairs of electrons of same spin are not possible to keep together since they repel each other. On a note, an analogous concept will later be discussed to put in effect the **Pauli exclusion principle**, which does not allow particles of same spin occupy the same energy state.

A question of interest is why electrons within atoms would interact through the magneto-motive potential (which has imaginary origin), and not simply through the real magnetic field (which has real origin). The answer is that since the wave nature of electrons has a real as well as an imaginary part, electrons may (as conditions allow) interact through both of those. It seems that the real part of the wave function interacts through the conventional magnetic field, while the imaginary parts interacts through the magneto-motive (so-called intrinsic) field version, taking advantage of the fact that two imaginary entities may interact and bring a negative real result. This point is of significance as it also directly associates to how charge itself forms inside atoms. An introductory approach on that topic is provided right next.

<u>Point 6</u>: How the magnetic field arises

According to textbook theory the magnetic field arises either by the motion of charge, or in respond to a change in the electric field (or flux). The case concerning a change in the electric field was considered above through the equations (5-4) and (5-6), while now we shall consider how the magnetic field arises in relation to motion, or effective motion of charge. For this, we have to consider two cases, ascribing to the particle nature and the wave nature of electrons respectively:

- The one case concerns the formation of the classical magnetic field, whose field lines form loops around the flow of electric charge. This seems to concern the motion of electrons' real portion (not the imaginary portion) which associates to the **particle nature** of electrons. In this section we shall consider the deeper origin of this version of the magnetic field via a reference to Ampere's law (5-3).

- The other case concerns the formation of the intrinsic magnetic field, which seems to have orthogonal properties with respect to the classical one, as its field lines were explained to be radial-like. This variant appears to relate to electrons' imaginary attributes, which do not involve motion in the conventional sense, instead they concern the **wave nature** of electrons. We shall start by considering the later:

Magnetic field as per the wave nature of electrons:
We shall here address how the wave nature of electrons takes essence inside atoms, without these particles' formation exhibiting conventional angular momentum. While the particulars of this waviness will be considered mainly in sections 7 and 8 through reference to the Schrödinger equation, we shall here provide a first conceptual approximation. For this, we shall seek to use a classical physics approximation describing a wave or alternation, to use that as a model to be based upon. Borrowing from material to follow, the actual wave function of a particle seems to account for alternations in the time domain and the space domain, which seems to resemble an alternation in the fastness of passing time and the stretchiness of space. Therefore, the equation of waviness or alternation that we would need to use as a reference base, should address such an alternation in time or space. As we are not accustomed to equations doing this, and would need to figure the parameters with respect to which this should be taking place. We may attempt to start by using a classical equation and make an **inversed sort of reading** of it. In particular, we shall refer to Maxwell's 4th law (5-4), the reason being that it involves the permittivity and permeability terms ε and μ, where these involve a shifting in time $\pm\Delta t$. For reasons of convenience we shall copy here Maxwell's 4th law, along with its modified expression (5-7) that we used earlier:

$$1/\mu_o \oint B \cdot d\ell = \varepsilon_o \, d\Phi_E/dt \qquad\qquad \text{(5-4) or (5-25)}$$

$$(1/\mu_o) \, dB / dx = (\varepsilon_o) \, dE / dt \qquad\qquad \text{(5-7) or (5-26)}$$

Let us focus on (5-26). Considering that the magnetic permeability is measured in Henries/meters (that is, L/dx) and the electric permittivity is measured in Farads per meters (that is, C/dx), we could imagine a different reading of this equation, where the right part may describe an oscillation in Capacitance C (instead of an oscillation of the electric field), and the left part of the equation may describe an oscillation in the reciprocal of Inductance L (instead of an oscillation of the magnetic field). That could loosely refer to the same relation now being red in the form of

$$B \, [d(1/\mu_o)/dx] = E \, [d(\varepsilon_o)/dt] \tag{5-27}$$

As we shall see later, these two different readings of that equation shall actually turn out to approximate the **key difference between the essence of bosons and the essence of fermions**. That is, instead of having oscillatory values of B and E and fixed values of inductance L and capacitance C (that applying for bosons), we come up (in loose approximation) with exactly the opposite; an oscillation of the values of L (that is, and oscillation in the phase shift in negative time) and the values of C (that is, an oscillation in the shift in positive time) where this is counterbalanced by fixed values of B and E (that applying for fermions), in association to a particle's fixed electric field and intrinsic magnetic moment.

Even though this reading of the above equations is oversimplified, it does serve as a rough indication of what holds for atomic electrons, where:

- The particle's matter seems to relate to an oscillation in time, with the particle's gravitational field having a counterbalancing role through a waviness in the metric of time (deployed over space).
- The particle's charge seems to relate to an oscillation in imaginary space, with the particle's electric field having a counterbalancing role through a waviness in the metric of imaginary space.
- The particle's intrinsic magnetic field seems to relate to a cyclic oscillation of negative time $-\Delta t$ (contained within the term μ of the left part of (5-27)) which applies as a (centripetal-like) potential that keeps the electron charge of an energy state in loop-like formation.
- A mechanical counterpart of the magnetic field which we shall address next, seems to relate to a cyclic oscillation of space that applies as a (centripetal-like) potential that keeps the electron's matter balanced in a loop-like energy state, which in turn allows for the representation of a particle's wave in the familiar lobe-like formation, instead of involving actual motion.

In relation to the above, the electron's intrinsic magnetic field seems to associate to the co-existence of the particle's imaginary part (what we could loosely refer to as "hidden pole"). Its action contributes to the particle's static magnetic properties, while it does not contribute to the creation of the classical magnetic field around the flow of conventional conduction type of current.

<u>Magnetic field of particle-like nature</u>:
The conventional magnetic field develops around a current carrying wire, that is, around charges that move at certain velocity dx/dt. We shall here recall a point that was stated earlier, that the electric potential has units which resemble the "reciprocal of velocity" (that is, seconds per meters). Rephrasing that in reverse, a measure of "seconds per meters" is perceived in the real world as the electric potential. On the basis of symmetry, a similar feat should apply in inverse, meaning that what we perceive as conventional *mechanical* motion dx/dt should be perceived as a potential with respect to an opposite domain where space and time have inversed roles. However, such an opposite domain refers to the frame of reference of charges, due to the orthogonality between mechanics and electrism, where time and space have been stated to swap roles. What this tells, is that mechanical velocity should be perceived as a potential from charges' reference frame. It turns out that such potential identifies to the magnetic one (in symmetry to the Bernoulli pressure developing around fluid flow in mechanics), where **the equipotential lines/surfaces of that potential refers to the magnetic field**.

While the above supposition is based on symmetry, it seems to comply to quantitative predictions of the laws of relativity applied on magnetic and electric fields as described in the Feynman's Lectures on Physics [14]. In particular, when a point charge is moving with velocity $v=dx/dt$ within the magnetic field created by an external wire, then, from the reference frame of the point charge, it's the external wire that is moving (while the point-charge considers itself to be stationary). The effective motion of the wire does not cause a relativistic effect directly on the charges that constitute the current of the wire. However, the moving wire itself is effectively slightly space-contracted by a factor of $1/(1-v^2/c^2)^{1/2}$, and this space contraction is equivalent to the same current (amount of charges) contained in a slightly shorter wire length, so that counts as an increase in the density of charge in the wire. Consequently, the point particle feels an effective additional electric potential and force (instead of a magnetic force). This electric force (as seen from the frame of reference of the point-charge) is exactly the same in magnitude to the magnetic force (as seen from the frame of reference of the current carrying wire). Consequently, **mechanical motion (like the motion of the wire) is perceived as a potential from the reference point of charges**.

As per the above, the motion of charge forms a potential, which may be approximated quantitatively either as a magnetic OR an electric potential, based in the frame of reference used, where both cases reduce to the same thing. Notice that this may be understood to be a **capacitive-like potential formed around a current**. This can be seen in figure 5-10, where the capacitive potential is represented by radial arrows around the current-carrying wire (thick black arrow). And the magnetic field (circular lines of B) represents equipotential surfaces/lines of that capacitive potential.

As long as the magnetic field forms symmetrically around a current carrying wire, this implies that the capacitive radial-like potential is equally distributed, so no net magnetic force applies. While, if an external influence exists, like an external magnetic field, then these magnetic field lines shift and become denser in the one side of the current and less dense in the other, exerting a sideways force (magnetic force) on the moving charge or current as described earlier.

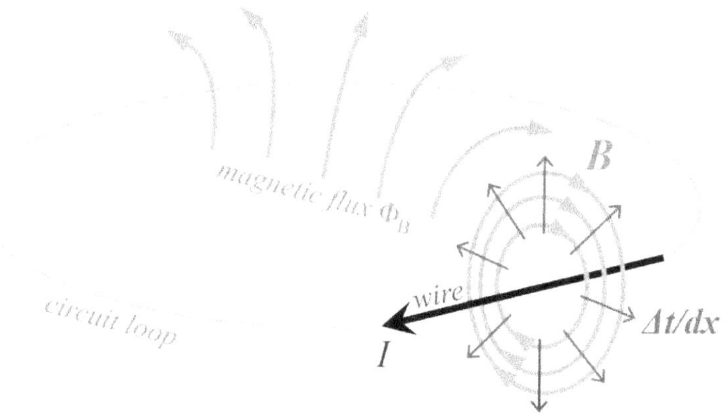

<u>Figure 5-10</u>: Current through a wire segment (thick black arrow), creates a capacitive potential $\Delta t/dx$ radially to the wire, the equipotential surfaces of which correspond to the magnetic field B.

While above we have been referring to the magnetic field of a straight wire segment, any current-carrying wire eventually forms a closed-circuit loop (so current may flow). In such case, the charges that flow in the closed circuit get to follow a loop-like path, which -for reasons of simplicity- we may imagine to approximate the motion of charges in a metal ring. For the charges of the wire to follow such a trail and move *circularly*, is physically similar to as free charges were moving through an external magnetic field and were being deflected sideways to move *circularly* (without change in their kinetic energy). In both cases of circular motion (either through the turns of the wire or due to magnetic deflection of a free moving charge) the charges end up moving as if a centripetal force is acting on them. In such a case, even though charges do not exhibit momentum (as mass does), the deflection corresponds to a sideways (centripetal-like) acceleration. For the charges that to move circularly by following the turns of the wire, the missing centripetal-like force reflects on (translates to) the value of inductance of the coil. If the current is changing, that comes along with a change in the radial-like potential $\Delta t/dx$ (and change in the associated magnetic field which corresponds to equipotential lines), and at the same time a corresponding counterbalancing effect of induction (an opposite phase shift in time) transmits a reaction to the electric source as explained earlier.

Notice that the change in the radial-like potential $\Delta t/dx$ resembles a change in reciprocal of velocity $1/v = 1/(dx/dt) = dt/dx$. From the point of view of a charge (considering that in electronics the parameters of time and space were stated to exchange roles with respect to mechanics) the corresponding change in $\Delta t/dx$ should be perceived as a shift or "flow" in the radial direction. That reflects a change in magnetic flux $\Phi_B = B\ dx^2$, which develops orthogonally (loop-like, around the wire) and should be considered an effect of imaginary character. That this is an effect of imaginary character may be more obvious by re-writing the same relation in the form shown in (5-28). The latter describes a "flow" of B *with respect to* an area of reciprocal space $(1/dx)^2$ rather than *times* an area dx^2. This is what accounts for the naming "flux" (through negative-reciprocal space), instead of conventional "flow".

$$\Phi_B = B\ dx^2 \quad \rightarrow \quad \Phi_B = B/(1/dx)^2 \qquad\qquad (5\text{-}28)$$

The magnetic flux may also be represented by the relation $\Phi_B = LI$, where L is the inductance and I is the current. That representation may address effects of reaction, when there is a change in current. It turns out that the magnetic flux tends to remain constant, pretty much like momentum tends to stay constant in mechanics. This may be justified by Lenz' law, according to which, if some externally imposed cause changes the magnetic flux Φ_B through a metal ring, a reaction current develops in the ring to counterbalance the imposed change in flux. It turns out that this applies as tendency for **the magnetic flux to be conserved** (instead of simply the *current* that produces the magnetic field to keep steady). We shall consider this point further in section 6.1.

Along the above lines, it is possible to rely on conservation of magnetic flux in deriving Ampere's law (5-3) via a pattern of calculation as follows.

$$\Phi_B = LI$$
$$B\ dx^2 = (-\Delta t)\ I$$
$$[dx/(-\Delta t)]\ B\ dx = I$$
$$[1/\mu_o] \oint B\cdot d\ell = I$$
$$\oint H\cdot d\ell = I \qquad\qquad (5\text{-}29)$$

In the next section we shall see how certain effects of electronics have a close analogy to effects of mechanics. For example, Lenz's law has a symmetry to the triggering of vortex flow (like in the whirling of water). Or, the radial-like capacitive potential $\Delta t/dx$ which develops around the flow of current has a close symmetry to the **Bernoulli** pressure. Relevant symmetries may extend in reference to negative forms of energy too, with one such case to be considered to explain spiral galaxies' rotation curves, without need for dark matter in the sense that existing theories hypothesize.

Capacitive and Inductive effects:

Effects of capacitance or inductance associate to a phase shift between voltage-current oscillations. These concern a phase shift of current either forward or backward in *time*. This shift provides to B and $\bar{B}$ fields a window to interact and leak energy toward the opposite energy domain in each case (imaginary or real) during transitional cases. That concerns effects as the following, which will be described in section 6:

- Effects involving a capacitive lead in time include the displacement of current between the plates of a capacitor, as well as a relevant process concerning the powering of electric batteries. These appear to involve the magnetic counterpart of the electromotive potential, which we'll be calling magneto-motive.
- Effects involving an inductive lag in time include self-induction, as well as the development of current in a metal ring when there is a change in flux through the ring, both of which involve the electromotive potential.

Note that analogous shifts and effects will be explained to develop in mechanics as well. The shifts in that case appear to develop in *space* (instead of time), as we shall figure right next.

5.4 The imaginary Gravitomagnetic Field (*T*)

Certain effects of fluid mechanics have symmetric characteristics with effects of electromagnetism, and based on that we shall explore how certain mechanical behaviors could be approximated quantitatively in a symmetric manner to their electronic counterparts. For doing so, a new, non-customary field needs to be introduced, which will allow to explain and to account for mechanical effects like the Bernoulli pressure around a fluid flow, vorticity, as well as to extend to effects of quantum nature as well.

We shall use the name **"gravitomagnetic"** for this field, for the sake of symmetry in the naming to the magnetic field. And we shall denote it with the letter T. Despite the similarity in the naming, the gravitomagnetic field does NOT concern a version of the known magnetic field. It relates to the gravitational field g in a symmetric way to as the magnetic field relates to the electric (hence the symmetry in the naming). In fact, it seems to have a subtle nature, as it appears to have imaginary character, and that gets it to have short range of action and quickly become evanescent in the classical world. To comprehend its nature and behavior, we shall go through certain paradigms to figure its involvement in effects which are happening around us.

We shall start by considering what holds with respect to side-pressure around the flow of a fluid, and shall explore a particular symmetry to what holds with the magnetic field around the flow of current. From the theory of fluid mechanics, we know that fluid flow associates to a low-pressure developing around the flow. This is demonstrated through the **Bernoulli principle**, which describes that the faster the flow of a fluid, the lesser the side-pressure at the walls of this fluid flow (figure 5-11).

The Bernoulli's principle applies on the basis of conservation of energy and the first law of thermodynamics, according to which, an increase in Kinetic energy (concerning the increase in the velocity of the fluid) comes along with a decrease in Potential energy (a decrease in side pressure on the walls of the pipe). This is plausible since energy can change form, while it cannot be created or destroyed. Other than this broad explanation on the basis of conservation of energy, however, there is no complete theory explaining how exactly the corresponding change in pressure takes place. The dispersion of molecules and atoms in a fluid is known to be attributed to Van der Waals forces, which are short-range forces that concern only the nearest particles, and they apply in an additive manner. However, not all components of these forces are understood to a full depth, and this includes the repulsive part which is considered to associate to the Pauli exclusion principle, addressed in section

5.3 point 5, as well as the attractive part which we shall be addressing in the present section.

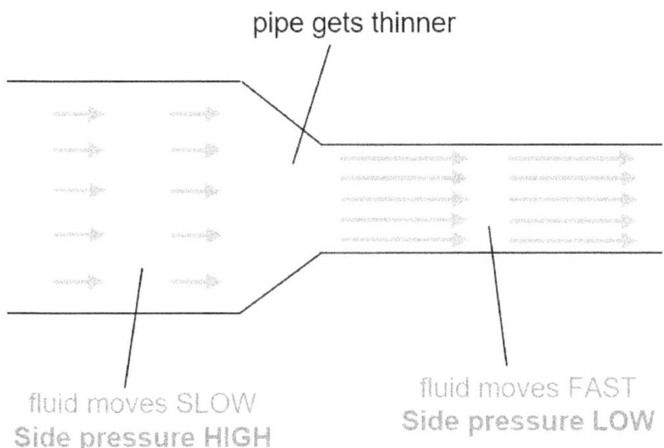

Figure 5-11: Fluid flow through a pipe that becomes thinner. The kinetic energy increases at the expense of side pressure

What appears to be the case, is that the Bernoulli pressure is driven by the gravitomagnetic field T and corresponding force. This field is actually coming in two versions.

- A classical physics version, which concerns the side-pressure effect that develops around fluid flow. Despite its origin being imaginary, we'll figure that the equations which describe it have certain symmetry to the equations describing the classical magnetic field, like Ampere's law (5-3) and Faraday's fourth law (5-4).
- A negative energy version that becomes involved in effects of quantum nature, and appears to concern a *mechanical* intrinsic "spin gravitomagnetic moment" of atomic particles, which becomes involved in exchange interactions between identical particles. This version of the field differs from the classical version in a similar way to as the intrinsic magnetic field was previously described to differ from the classical magnetic field.

We shall start by considering the classical version of the T field, which relates to the side pressure around a fluid flow. That pressure is more obvious and easier to measure when a fluid flows through a pipe (because, once the fluid comes out of the pipe, fluid particles may move lateral to the flow in which case the sideways pressure is less obvious). Note that according to material presented later in this section, the side pressure should *not* be considered limited to motion of fluids. In principle it should apply to motion in solid form too, however in such case the side-pressure does not exhibit itself because molecules are bound to the solid so they have no freedom to disperse away in order to exhibit side pressure in a collective manner.

When a fluid flows through a pipe, the side pressure that develops on the walls of the pipe has some symmetry to the radial-like capacitive differential in time ($\Delta t/dx$) that was previously described to develop around an electric current flowing through a wire. As it was explained then, the equipotential surfaces of such potential correspond to what we perceive as the magnetic field (see figure 5-12, right part). It turns out that **a symmetric condition holds in the case of the flow of matter in fluid form, which comes along with a radial-like field which has the parameters of space and time inversed so it corresponds to a differential of the form $\Delta x/dt$. This also applies as a potential, and corresponds to the potential that drives the Bernoulli pressure, of which the "equipotential-pressure" lines (or surfaces) correspond to the gravitomagnetic field T.** A schematic illustration of the radial-like, Bernoulli pressure-related potential is shown in the left part of figure 5-12 (see radial arrows $\Delta x/dt$), in visual comparison with the potential of the magnetic case (same figure, right side).

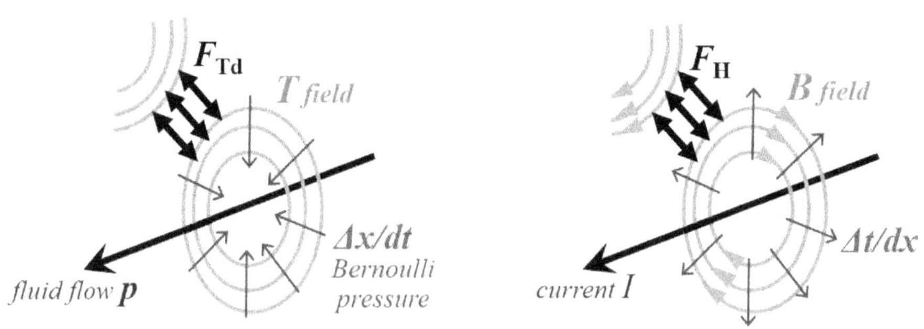

Figure 5-12: Schematic symmetry between
magnetic field B & displacement magnetic force F_H (right),
and gravitomagnetic field T & displ. gravitomagnetic force F_{Td}
which concern the Bernoulli pressure (left)

The swapping of the parameters of space and time between the gravitomagnetic and magnetic cases, goes along with the following differences in behavior:

- In the magnetic case, the density of the magnetic field lines seems to concern differentials in imaginary (negative reciprocal) time, which seem to interact with the imaginary part of the particle's wave function. In such case, due to the involvement of two imaginary entities, we end up with the magnetic action bringing an outcome in the **real** world (due to the relation $i^2=-1$). And since the outcome is *real*, the magnetic interaction may apply over long range (meaning that it does not damp out in very short range).

- A different condition holds for the gravitomagnetic case. As per the orthogonality between matter and charge depicted in figure 4-1, the gravitomagnetic field has an imaginary character, relating to a differential in space contraction. This means that the corresponding potential should provide its effects in negative energy (in the imaginary domain), and because of that the gravitomagnetic field becomes quickly evanescent in the real world, so it damps out at *short range*. This is why it may produce real results (pressure-related) only very close to the edge of the fluid flow. In relation to that feat, the equipotential surfaces concerning the T field lines shown in figure 5-12 left part (concentric grey loops) are only schematic for illustration purposes, while the actual field would only apply at a very short range upon the edge of the fluid flow. Furthermore, these field lines are **direction-degenerate**, meaning that they do not have a preferred direction of turn. Either direction of turn works equally well, this is why the T field lines in figure 5-12 are not shown to have an arrow indicating direction). Actually, the underlying reason for this degeneracy seems to be that only a single sign of matter makes it to the real side of the universe (since both matter and antimatter constitute "positive matter", as they are both gravitationally attractive).

In the absence of any external gravitomagnetic field in the neighborhood of a fluid flow, the T field lines are equally distributed so they form co-axial equipotential surfaces. And as long as they stay co-axial there is no net gravitomagnetic force applied. While if an external influence gets to deform the evenly distributed shape and alter the density of these T field lines, this change in density identifies to a potential and a corresponding gravitomagnetic sideways force (in symmetry to as the magnetic force was explained to apply in the previous section). One example at a collective macroscopic level, concerns the sideways displacement of tornados. Actually, while tornados are a large-scale effect, the *short-range* gravitomagnetic potential may still bring result due to the involvement of a very large number of adjacent molecules that mediate the effect in a collective manner.

The above concepts bring up a number of points of attention, as follows:

Point 1: Gravitomagnetic force action and reaction

In relation to the inverse symmetry between the gravitational and electromagnetic interactions which involves a swapping between the parameters of space and time, it appears that the gravitomagnetic field T relates to the gravitational field g as per the following relation:

$$g = (1/\upsilon)\, T \qquad\qquad (5\text{-}30)$$

In a comparison between this equation and its electromagnetic counterpart $E=\upsilon B$, one may notice that the velocity term is reciprocated, which is due to this swapping of the parameters of space and time. In fact, since the velocity term here has moved to the denominator, this is what **gets the Bernoulli side pressure be smaller, the higher the fluid velocity**, as shown in figure 5-11.

The particular way that the T field may get to affect its force on matter resembles the process of the magnetic action. And just like the conventional magnetic force $F=q(\upsilon xB)$ has some symmetry to the electric force $F=qE$, a similar kind of symmetry should be expected for the mechanical case, between the gravitomagnetic force and the gravitational force $F=mg$. In such case the relation (5-30) provides the hint that in the gravitomagnetic case the "force agent" seems to be the combined term $((1/\upsilon)xT)$, and not the gravitomagnetic field T by its own. This is like saying that what we may call gravitomagnetic force is not a force of T, it is rather a force of $((1/\upsilon)xT)$, where the corresponding relation describing this force, is

$$F = m\,(1/\upsilon \times T) \qquad\qquad (5\text{-}31)$$

While this equation may be fine for the present purposes, it is worth noticing that it should be more appropriate to replace "mass" by a term that would reflect "matter". This is so, because mass is rather a relative quantity which may be subject to influences (for example, effects of relativity), while matter in the other hand, is a more absolute quantity (just like charge is a more absolute quantity, in electronics). In order to reach a formulation that would rely on matter, one should involve the Planck constant, which is a "constant" as its name implies. The transformation could be achieved by equating the classical momentum ($p=m\upsilon$) with the momentum as per the wave nature ($p=h/\lambda$), so we get $m\upsilon=h/\lambda$, and from that

$$m = (h/\upsilon\lambda) \qquad\qquad (5\text{-}32)$$

So, the term m in (5-31) could be substituted by (5-32). But then, if we consider that for energy it holds $E=hv=h(v/\lambda)$, one could even replace the mass term m with is equivalent E/c^2 where c is the speed of light. In that case (5-31) would be expressed as

$$F_{(1/\upsilon \times T)} = (h/\upsilon\lambda)\,(1/\upsilon \times T) =\ (E/c^2)\,(1/\upsilon \times T) \tag{5-33}$$

We shall now check if the above relations comprise an accurate description of how the corresponding force applies, or, if it would require some fine-tuning. To do this, we should carefully look into an example where this force applies. To find a proper example, we should keep in mind that we are looking for an interaction that
- causes a sideways deflection force,
- it should apply to moving matter (in particular fluid flow), and
- it should affect the deflection of matter without changing its kinetic energy (much like it holds with a moving charge that is being deflected by an external magnetic field).

One may consider whether a proper example for the above, could concern the involvement of the gravitomagnetic force in an effect of vorticity. As we shall figure in section 6, vorticity typically arises via a different type of interaction. While instead, in the present case we need to realize that the force we are looking for only concerns a sideways deflection, and should therefore be a RADIAL-like (centripetal-like) force, and NOT a loop-like potential (where the latter will be associated to the triggering of vortex flow, which is a different effect). In order to avoid confusion, we shall present the gravitomagnetic force action in a rather plain and familiar embodiment, that of the lift force of an airplane wing via the Bernoulli principle, illustrated in figure 5-13.

In this case, instead of having a fluid to flow through a pipe as in figure 5-11, we have a material object (an airplane's wing) moving through air with velocity υ. Notice that in this example the "angle of attack" of the wing is zero, so the upward lift is not at all attributed to the inclination of the wing, it is attributed solely to the Bernoulli principle.

As per common knowledge, the principle behind the lift has to do with that the upper part of the wing is more curved, which makes air travel a larger path in the upper part of the wing than it does under the wing. This in turn makes air move faster in the upper side of the wing than in the lower, and the faster flow exhibits lower side-pressure according to the Bernoulli principle.

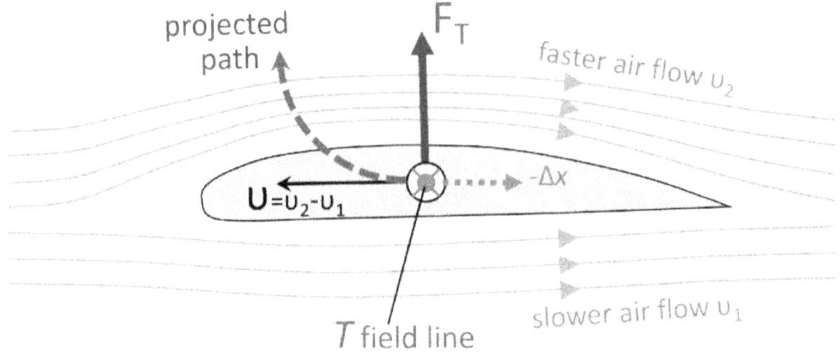

Figure 5-13: Airplane wing's lift principle. The gravitomagnetic field T is perpendicular to this page, and is degenerate in direction

So, the lower side-pressure at the upper part of the wing, along with the higher side-pressure at the lower part of the wing, create the lift force. But let us seek a deeper insight on how this takes place.

<u>A. The Action part of the gravitomagnetic interaction.</u>
To figure how the gravitomagnetic force may lift an airplane's wing, we may try to apply a sort of a "left-hand-rule". Actually, in this particular case it does not matter if it is a left-hand or a right-hand rule, because the field itself is direction degenerate (indifferent). What matters, is to carefully identify which interacting entities are involved in each of the 3-axes which are orthogonal to each other.

The T field's lines may be corresponded to **equi-pressure** lines (in symmetry to as the magnetic field lines were explained to correspond to equi-potential lines). In this respect, the direction of the T field lines may be thought of to identify to the wing itself (the direction perpendicular to the page). This is so, since air passing above the wing has lesser side pressure, while the air passing below the wing has higher side pressure, so the wing itself roughly corresponds to an equipotential surface.

A counterintuitive point that the existing theory for the lift does not address, is that, since the upper part of the wing is more curved, air flowing above the wing should be expected to be more compressed (in the direction perpendicular to the flow) and this should rather be expected to "push down" the wing instead of "lift it up". The explanation to this tricky point may be traced in (5-31), in the reciprocation of the velocity term. Remember that the gravitomagnetic force is not a force of T, it is rather a force of $((1/v)\text{x}T)$. So, in accordance to that, the *faster* the velocity v of the air, the *weaker* the gravitomagnetic force, thus the less the side-pressure at the upper part of the wing, hence the lift.

This explanation may actually go a step further, as follows: That the upper part of the wing is more curved, gets the (faster moving) air molecules above the wing to get slightly more compressed and therefore get shifted marginally closer to each other (in the direction perpendicular to the air flow). This may be seen in figure 5-13 where the air flow lines (and corresponding density of equi-pressure surfaces) are much closer to each other in the upper part of the wind than in the lower part of it. (Notice how symmetric this is, to the change in density of the lines of a magnetic field when a charge moves through it). This change in proximity of the T field lines addresses a **Van der Waals** potential. In this case, as molecules are getting closer to each other, this applies as a negative shift in space Δx. And since this shift in space develops with respect to time, it may be accounted for by a term $\Delta x/dt$, which corresponds to a "capacitive-like" differential, and potential.   (Notice how inversely symmetric this term is to the capacitive potential $\Delta t/dx$ in the case of the magnetic deflection force, described earlier).

A point of attention, is that in the case of the magnetic force we were dealing with a change in the density of field lines only (not some change in density of moving *charges*). While in the mechanical case, in the contrary, we have a change in the density of *matter* in fluid form (not a change in the density of lines of a field). This however, does not seem to constitute a much different case. It has to do with the orthogonality between matter and charge (shown in figure 4-1), according to which, the gravitomagnetic field has imaginary essence as stated earlier. This imaginary attribute has the consequence that **its displacement shift concerns the displacement of matter** (this is in symmetry to as the displacement of electric flux displaces charge between the plates of a capacitor). This is actually how the very large number of co-involved molecules engages in a collective (transmissive) manner, and gets the gravitomagnetic force to bring result at a relatively large scale.

The units of the differential $\Delta x/dt$ (if accounted for in meters per seconds) may look similar to those of the conventional velocity $v=dx/dt$, but they represent something completely different, they represent a "*phase shift* in space $\Delta x/dt$". Such term $\Delta x/dt$ may apply as a   permeability-like term μ_g (the subscript "g" referring to the "gravitational" context, which differentiates it from its magnetic counterpart). We may call that term "gravitomagnetic permeability", or "**mechano-permeability**". Despite the similarity in the naming, this is fundamentally different to the magnetic permeability μ_o which is measured in Henries per meter, where a Henry is the measure of inductance, and inductance corresponds to a shift in the **time** metric as described previously. In the mechanical case instead, the permeability-like regulator μ_g represents a similar type of phase shift, but here the shift is in the **space** metric (as in the change in proximity of the gravitomagnetic field lines illustrated in figure 5-14 right part). We could name the units of the shift in space **"mechano-Henries"**. The shift, in turn, could be referred to as mechano-inductive and may be denoted as L_g (the subscript "g" referring to the "gravitational" context, once again). So, a shift L_g with respect to time dt, yields the aforementioned mechano-permeability term $\mu_\mathrm{g} = -L_\mathrm{g}/dt$.

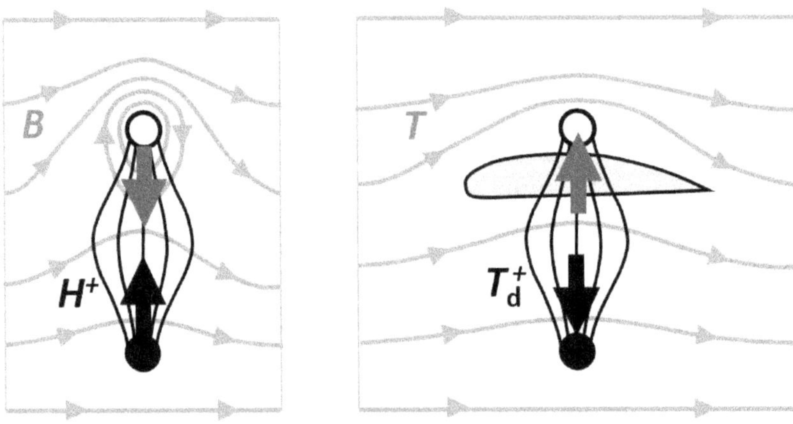

Figure 5-14: Comparison between magnetic force on moving charge (left) and gravitomagnetic force lifting a wing (right).

Notice that through this process, the curvature of the wing (across its length which extends "inside the page") as depicted in figure 5-14 right part, plays a similar role to that of the motion of current (figure 5-14 left part, where the white circle represents a charge moving inside the page) which shifts the lines of the magnetic field, and therefore creates a potential $H+$ which in turn deflects the moving charge. Notice that the left part of figure 5-14 illustrates the same effect as in the right part of figure 5-7, rotated anticlockwise by 90°. Here, one may notice that the shifting of the magnetic field lines by the moving charge (white circle) deflects the charge down. The right side of 5-14 exhibits the similar effect in mechanics, where the air above the wing is denser (the white circle resembling a hypothetical pole), and this identifies to a potential $-\Delta x/dt$ which applies a sideways force to the wing, where in this case the force points in the **opposite direction** (outwards) lifting the wing up.

In relation to all the above, the difference in the density of the gravitomagnetic field lines above and below the wing correspond to a difference in contraction of the space metric. Notice that this is pretty symmetric to the difference in the density of the magnetic field lines in figure 5-7 left part. Now, the role of the "external magnetic field" has been taken over by the density of the air coming from the front of the wing, and the role of the "moving charge which deforms the external magnetic field" has been taken over by the curvature of the wing (across its length which goes inside the page). The interaction of the air (coming from ahead) with the curvature of the wing creates the deformation of density of gravitomagnetic field lines, represented by the term μ_g which reflects a potential. That combines with the T field to create a new displacement field which we shall call **displacement gravitomagnetic field**, and denote it by the symbol T_d where the subscript "d" reflects its displacement nature. This field corresponds to the following relation, where the mechano-permeability has been reciprocated because it reflects negative energy, so it has imaginary nature.

$$T_d = (1/\mu_g)\, T \qquad \text{[approximation]} \qquad (5\text{-}34)$$

The above relation however is not fully accurate. The reason is, that T_d includes the mechano-permeability μ_g, and that in turn includes mechano-inductance L_g which corresponds to a negative shift in space $-\Delta x$, but if such a shift were indeed inductive-like, it should develop loop-like (similar to as an inductive phase lag does over an electric circuit loop). While in the present case we have a differential in space which develops radial-like, as it has a capacitive character, with a tendency to displace the wing sideways-upwards (ideally in symmetry to as a moving charge is deflected to move circularly by a magnetic field), see the dashed curve of the "projected path" in figure 5-13.

In such case, we shall address the existence of TWO types of displacement gravitomagnetic fields, both of which have similar raw units, but they have opposite signs. We shall denote them by the symbols T_d^- and T_d^+, where the superscript (−) refers to an "Inductive displacement Gravitomagnetic field", and the superscript (+) refers to a "Capacitive displacement Gravitomagnetic field". These ascribe to the following relations:

$$T_d^- = (-1/\mu_g)\, T \qquad\qquad (5\text{-}35)$$

$$T_d^+ = (-1/\varepsilon_g)\, T \qquad\qquad (5\text{-}36)$$

Here, the μ_g corresponds to $-\Delta x/dt$ while ε_g corresponds to $+\Delta x/dt$, so they are seemingly similar entities that differ in the sign of the phase shift. Their opposite sign also implies that they operate in opposite energies, which suggests that either one of the two should operate in imaginary conditions with respect to the other.

Once again, despite the similarity, ε_g is substantially different to the electric permeability ε_o which is measured in Farads per meter, where the Farad is the measure of capacitance, and capacitance corresponds to a shift forward in the **time** metric as described previously. In the mechanical case instead, the permittivity-like regulator ε_g represents a similar type of phase shift, but now the shift is in the **space** metric (relating to the proximity of the gravitomagnetic field lines). We could name the units of the corresponding entity **mechano-Farads**.

So μ_g and ε_g are similar entities that seemingly differ in the sign. That in turn signifies that only one of them, either μ_g or ε_g should be in reciprocal form. That is, since the mechano-permeability comes in reciprocal form $-1/\mu_g$, the permittivity should appear in ordinary from ε_g. However, it seems that ε_g becomes involved in an interaction in opposite energy. (More will be said on this in the next section. It is like saying that the former interacts with the real portion of particles' wave nature, while the later interacts with the imaginary portion of particles' wave nature). Hence, this interaction should involve the mechano-permittivity in reciprocal form too, as the term (5-36) describes. Due to this, both (5-35) and (5-36) involve the mechano-permeability and mechano-permittivity terms in reciprocal forms. In such case, **the conventional velocity $v=dx/dt$ of a matter in fluid form creates an opposite potential $+\Delta t/dx$ as viewed with respect to opposite energy**. Consequently, the relation (5-31) may be re-written more accurately as follows, where the notation of the displacement gravitomagnetic field includes the positive mark. (Notice also, that in such case the term $(1/\varepsilon_g)$ points in orthogonal direction to $(1/v)$).

$$F = m \, T_d^+ = m \, ((1/\varepsilon_g) \, T) \tag{5-37}$$

At this point it is worth to highlight the relative resemblance between the displacement gravitomagnetic field T_d^+ and the gravitational field g, which relate to each other much like H^+ relates to E in the electromagnetic case. That is, T_d^+ and g resemble each other to the same extend as $((1/v)\times T)$ resembles $((1/v)T)$. Therefore, the force that we called gravitomagnetic actually concerns a **"Capacitive Displacement Gravitomagnetic"** potential, and the way it acts is by involving a mechano-capacitive shift which is accounted within ε_g. This potential deflects the wing's direction of motion upward without changing its kinetic energy, in symmetry to as a magnetic field deflects sideways a moving charge to follow a circular-like path (see "projected path" in figure 5-13). Of course, in the case of mechanics other co-involved forces (like gravity and frictional forces) prohibit this path from having a circular shape.

B. The Reaction part of the gravitomagnetic interaction.
The upward deflection (lift) of the wing corresponds to a sideways acceleration and that applies as a mechanical "load". In respond to doing work in accelerating that load, some energy needs to be retrieved from somewhere else, to pay back for the work done.

To remind what holds in the magnetic case, getting electrons to circle through a coil's spiral associates to the value of inductance of the coil, and thus a change in current becomes subject to an inductive phase shift in time, through which the system borrows energy through the time domain. This is perceived by us via a phase lag in Current with respect to Voltage, where this lag conveys negative energy to the electric source that powers the charge's motion. In quantitative terms we may consider that this reaction process leaks energy from the real domain toward the imaginary domain (since part of the energy provided gets consumed without doing work), thus reducing the efficiency in delivering actual work (for example, reducing the efficiency of an electric engine) due to the associated impedance. This part of the energy is represented through a Sine coordinate.

A symmetric effect of reaction takes place in mechanics, which also involves negative energy (an imaginary coordinate), thus the reaction takes place in an orthogonal direction with respect to the action force. In particular, this comes as a mechano-inductive effect developing in space, which corresponds to a drag in space (instead of a lag in time), and that identifies to a potential that drags the airplane backwards, impeding its acceleration. (See conceptual depiction of the dragging shift $-\Delta x$ in figure 5-13). Through this way, the drag borrows energy from the source which provides the energy for the motion of the airplane (which is the airplane's engines). Same as in its electronic counterpart, this borrowing of energy seems to apply as a Sine coordinate of energy, so it applies as a coefficient of reduced efficiency. Consequently, **only a Cosine coordinate of the energy supplied by the plane's engine delivers actual work** (provides thrust in the real domain) while the Sine coordinate corresponds to the portion of supplied energy (from the plane's engine) that gets consumed in exercising the drag (shift backward-in-space). While this drag does not deliver actual work, it leaks energy toward the imaginary domain and in this way reduces the efficiency in obtaining thrust, in return to the lift.

The overall transitional effect of the airplane wing's deflection upward during takeoff, operates in a similar way to a change in current through a coil, in which case:

- In the magnetic case, an inductive displacement potential (a lag) conveys negative energy (in the time domain) to the battery, which translates to reduced efficiency in accelerating charges.
- In the gravitomagnetic case, a mechano-inductive displacement potential (drag) conveys negative energy (in the space domain) to the airplane's engine, which translates to reduced efficiency in providing thrust to the airplane.

And just like in the magnetic case the reaction potential is associated with a version of the electric field which has loop-like field lines and drives the electromotive potential, in the gravitomagnetic case the reaction potential is associated with loop-like version of the gravitational field, which we may call **gravitomotive**. The gravitomotive potential impedes the acceleration of the wing and airplane, while in the next section we shall figure that such type of potential also becomes involved in another orthogonal interaction which drives effects of vorticity. The energy provided by the source (the airplane's engine) becomes subject to an inefficiency due to a sine coordinate of an angle φ involved, where that angle reflects the associated phase shift between the Space and Time metrics (generated due to the wing's curvature), causing the drag.

In an actual airplane measures are taken to change the lift force. This is done by affecting the curvature in the upper part of the wing, to have it become more extensive and pronounce. This is actually exercised in airplanes during takeoff and landing, by extending the wing's flaps (at the back side of the wing) and the slats (at the front side of the wing) so as to enhance the overall curvature in the upper part of the wing.

Furthermore, just like the denser (stretched-in) gravitomagnetic field lines at the upper part of the wing deflect it upwards, the more far apart (stretched-out) field lines below the wing exert an opposite force on that layer of air below the wing, pulling that layer of air down, toward the denser air layers existing further below it, as shown in figure 5-14, right part.

A point worth noticing, is that in the above paradigm with the airplane wing, the T field lines were suggested to nearly coincide with the wing itself (from the body of the airplane until the tip of the wing), since that direction concerns almost equi-pressure lines/surface. However, since the T field lines concern to equipotential lines, they should normally form loops (just like the magnetic field lines form loops, too). In this case, the straight portion of the T field lines (parallel to the wing) can be thought of to tend to become part of larger hypothetical loop, where this tendency relates in some degree to the formation of vortexes of air at the tip of the wing.

The above interaction could be put in the form of an equation (as for the magnetic case too). The one side of the equation could reflect the change in density of the gravitomagnetic field "lines" (a potential created as a result of the air flow facing the curved wing). The other side of the equation could describe the wing's deflection (lift) as a form of a centripetal-like acceleration (a "sideways flow") of matter, which applies radial-like. A corresponding equation that describes the gravitomagnetic force should look much like Faraday's fourth law in reverse order. This reverse order also implies an orthogonality. The change in gravitomagnetic field is now radial-like (sideways to the direction of the gravitomagnetic field lines), and the deflection of the wing concerns actual motion of matter (instead of a "displacement momentum", with more to be said on that ahead).

<u>Point 2</u>: Raw Units of the gravitomagnetic field

To figure out the "**raw units**" of the gravitomagnetic field, we may proceed similarly to as it was done previously for the magnetic field. For that, we shall rely on the relation (5-30) between the gravitational and the gravitomagnetic fields, which reads $g=(1/v)T$, as well as on the effective units of the gravitational field dx/dt^2. In such case, the units of T should refer to $T=vg$ which corresponds to $[(dx/dt)(dx/dt^2)]$. That translates to (m/sec)(m/sec^2), which eventually makes for **m^2/sec^3**. Notice that this is symmetric to the raw units of the magnetic field being sec^2/m^3, subject to the swapping of the parameters of space and time.

Here again, these specific units by their own do not seem to reflect an "acceleration term" that would justify causing a force (like for example the gravitational force resembles the acceleration term dx/dt^2, or like the electric force involves a space-to-time swapped "acceleration-like" term $dt/d(-x)^2$). An acceleration term may however be found in the **displacement gravitomagnetic field** T_d (either the capacitive $T_d^+=(-1/\varepsilon_g)T$ or the inductive $T_d^-=(-1/\mu_g)T$ displacement version) whose units concern $(1/(\Delta x/dt))\ (dx^2/dt^3)$ which makes for **m/sec^2**. These units resemble the units of acceleration, and that may justify that the displacement gravitomagnetic field acts as the agent of the gravitomagnetic action force, in compliance to the aforementioned concepts.

A point of significance is that the units of T that were presented above may also be re-written differently, by splitting the original term (m^2/sec^3) to **(m^2/sec)(1/sec^2)**, which may alternatively be expressed in the form $(dx^2/dt)(1/dt^2)$. The latter includes two terms; an acceleration-like term (dx^2/dt) as well as a flux-like term $(1/dt^2)$. This expression indicates an indirect complementarity between the gravitational and electromagnetic interactions, as the former term actually resembles a negative reciprocation of the raw units of the **electric** field $d(1/t)/d(1/x)^2=(dx^2/dt)$, while the latter term describes a **flux** in time. So, in a sense, the units of the gravitomagnetic field T seem to also correspond to a flux of anti-electric field $\Phi\bar{E}$.

$$T \sim \bar{E}\,(1/dt^2) = \Phi\bar{E} \qquad\qquad (5\text{-}38)$$

Point 3: Maxwell-like equations of gravitomagnetic wave propagation

While the above concerns the T field developing around a steady fluid flow, we shall now explore this field's behavior in transitional effects which may include changes in the gravitational field g. This can be addressed through the following relations which are symmetric to Maxwell's relations subject to a swapping between the parameters of space and time (since such swapping is here discussed to hold generally between the gravitational and electric interactions). In relation to that, equation (5-39) describes a change of a quantity that we may call **gravitomagnetic flux**, with respect to *space* dx (instead of with respect to *time* dt which was the case for the magnetic flux in equation (5-5)). And it similarly applies for relation (5-40) which is symmetric to Maxwell' fourth law (5-4).

$$\oint g\cdot d\ell_t = -\,d\Phi_T/dx \qquad\qquad (5\text{-}39)$$

$$(1/\mu_g)\oint T\cdot d\ell_t =(\varepsilon_g)\,d\Phi_g/dx \qquad\qquad (5\text{-}40)$$

Equation (5-39) describes a reaction process through an effect which we shall call "**gravitomotive**" and will be explained to be responsible for the vorticity of fluids. The right side describes a change in gravitomagnetic FLUX, so it concerns a flux of the T field along the direction of the field arrows (either forward or backward, but not sideways). The left side describes a negative energy (imaginary) version of the gravitational field, which has loop-like field lines (instead of conventional radial-like field lines), and will be explained to relate to a loop-like displacement of matter in effects like the whirling of water (in symmetry to the displacement of current in a metal ring when there is a change in magnetic flux through the ring). The term $d\ell_t$ concerns a particular loop in the time domain, and we shall loosely handle it in units of "dt" (that is, seconds). We will deal further with the very particulars of that interaction in section 6.2.

One may re-arrange the terms of (5-39) by substituting the gravitomagnetic flux Φ_T with an expression which we shall postulate to hold on the basis of symmetry to its magnetic counterpart $\Phi_B = B\, dx^2$. According to such symmetry, for the gravitomagnetic flux Φ_T it should hold:

$$\Phi_T = T\, dt^2 \qquad\qquad\qquad (5\text{-}41)$$

This expression may seem to run into an intuitive issue. Notice that in the case of the magnetic flux the term "dx^2" refers to flow through **an area**, as it uses two dimensions of space (out of the three dimensions of space). If a similar concept is adopted for the gravitomagnetic case, the term "dt^2" could **assume the "use" of two dimensions of time**. If so, that would raise a contradiction with respect to current theory which accepts a single dimension for time (the so-called "fourth" dimension on top of the three dimensions of space). If time had one dimension only, then the term dt^2 in the expression of Φ_T could rather stand improper. We shall here consider the possibility that the conventional notion that time has only one dimension should not be accurate. That can be figured through a reference to a thought experiment:

Consider two travelers moving in independent rockets, in parallel paths side-by-side to each other, at a relativistic velocity. For both of them, time is more dilated with respect to the time experienced by people on earth which they have left behind. However, time is not delated with respect to each other (as their rockets are moving side-by-side to each other). In that sense they experience more than one dimensions of time; one with respect to time on earth, and another with respect to each other. For conceptual reasons only, it may seem helpful to associate dimensions of time to dimensions of space so as to easier conceive how each dimension is independent of the others. For example, by assuming that, if one dimension of time associates to the direction of motion, the other two could be along perpendicular directions. However, the dimensions of time should *not* actually stick to dimensions of space. In the above paradigm involving two travelers, for instance, the travelers do not need to move in parallel paths in space, as the same effect would hold if each of them travelled along different directions. A similar thought experiment could be extended to include another traveler, experiencing different time dilation, where that may allow to speak for **three dimensions of time**. Thus, an entity like a human person may experience independent coordinates toward each of these dimensions of time. For instance, physical acceleration could affect only one dimension of time, leaving the other two dimensions unaffected. But this may also depend on the chosen reference frame. Concurrently, a second or third dimension of time could be put in use in accounting for effects concerning inductive or capacitive shifts (in association to particles' charge).

So, according to the above, it should be considered okay for the relation of the gravitomagnetic flux $\Phi_T = T\,dt^2$ to use two dimensions of time in (dt^2). In fact, in the particular case the multiplication of T with dt^2 would be more proper to express as a division by ($1/dt^2$) which is mathematical equivalent. This corresponds to having a flux of T with respect to an "imaginary-time-area" of ($1/dt$)($1/dt$). This may seem as a counterintuitive notion, but it should be fine to use. Therefore, as per the above we may accept for (5-41) to be valid, and we may proceed to plug it in to (5-39), in which case that loosely transforms to $g\,dt = -\,T\,dt^2/dx$, and then to:

$$dg/dt = -\,dT/dx \tag{5-42}$$

This transformation is similar to the one that was done earlier, where from equation (5-5) we had arrived to (5-6) which concerns the propagation of an EMW. Similarly to as was the case then, the transformation to (5-42) should be considered somewhat loose as this equation describes an interaction where the fields traverse SIDEWAYS to the direction of the field lines. While instead, the source equation (5-39) concerns an orthogonal case where the change takes place along the direction of the field lines of g. In any case, equation (5-42) may be considered to address the propagation of a travelling gravitomagnetic wave, which takes place in a symmetric way to as an EMW is conventionally considered to propagate. Actually, for reasons of symmetry in the naming, that wave would be more accurate to call "gravito-gravitomagnetic" wave. But for simplicity we shall call it "gravitomagnetic" wave, or GMW.

Notice that while g oscillates per *time*, the T oscillates per *space*, and that signifies that the two fields have counterbalancing roles in complementary domains (one in space and the other in time), therefore either of them should be imaginary with respect to the other. This also complies to the fact that they deploy orthogonally to each other.

Furthermore, a re-arrangement of the terms in (5-42) may lead to $g = -(dt/dx)T$ which matches the aforementioned relation $g = (1/\upsilon)T$. Notice how symmetric that is to $E = \upsilon B$ (or rather, how antisymmetric to it, due to the reciprocation of velocity). In any case, the relation (5-42) does not contribute in specifying the value of the velocity that would allow a GMW to propagate without impedance.

We shall then focus in the relation (5-40), and re-arrange its terms just like we did previously with (5-39), by replacing the gravitational flux Φ_g with an expression which we shall for now postulate to hold on the basis of symmetry as well, just as we did for the gravitomagnetic flux (and we shall support that further in the next section). That is,

$$\Phi_g = - g \, dt^2 \tag{5-43}$$

Based on that, (5-40) may loosely transform to $(1/\mu_g) \, T \, dt = \varepsilon_g \, g \, dt^2/dx$, and from there to

$$-(1/\mu_g) \, \boldsymbol{dT/dt} = (\varepsilon_g) \, dg/dx \tag{5-44}$$

Here again, the latter equation seems to describe SIDEWAYS propagation, instead of a change in gravitational FLUX which the parent equation describes and concerns a change along the direction of the field arrows. But other than this, equation (5-44) seems to provide more information about the propagation of gravitational waves than what (5-42) does.

A comparison between (5-42) and (5-44) reveals that a swapping has taken place between the parameters of space (dx) and time (dt) in the denominators of the g and T fields in these two equations. Other than the immediate reading of this swapping, there can also be a complementary reading by re-arranging the denominators of (5-44) so as to get the mathematically equivalent reading shown in (5-45)

$$(1/\mu_g) \, dT \, / \, (1/dx) = -(\varepsilon_g) \, dg \, / \, (1/dt) \tag{5-45}$$

$$dT/dx \quad = \quad -dg/dt \tag{5-46}$$

Now the left-side accounts for a change in T with respect to **imaginary space** $idx=(-1/dx)$, and the right-side accounts for a change in g with respect to **imaginary time** $idt=(-1/dt)$, where the two minus signs cancel out. Notice how this compares with (5-42) which we have re-written here in opposite direction (5-46), whose one side accounts for a change in T with respect to **real space** (dx), and the other side accounts for a change in g with respect to **real time** (dt). What this tells, is that (5-45) and (5-46) seem to approximate the same type of interaction from the point of view of opposite energy. Let us keep that in mind as it shall be useful in conceiving effects of reaction that will be considered in section 6.

In relevance to the above, a comparison between (5-42) with (5-44) also reveals that a swapping from real to imaginary coordinates of space and time in the denominators, necessitates the involvement of a mechano-permittivity (ε_g) and a mechano-permeability (μ_g) terms in (5-44), for maintaining equivalence. Or, to say the same in a different way, to measure T with respect to imaginary space $(1/dx)$ instead of real space (dx) requires the involvement of $(1/\mu_g)$, and to measure g with respect to imaginary time $(1/dt)$ instead of real time (dt) requires the involvement of ε_g. Therefore, in a way, **the involvement of the regulators ε_g and μ_g acts as a bridge between the negative energy and the positive energy domains**. (Notice that we have used the term "regulators" instead of "constants", since each of those may take different value in different cosmic regions and interaction particulars).

It further appears, that the role of ε_g and μ_g is not that of a simple "adjustment factor", it is instead a far more fundamental and subtler role. As mentioned previously, the μ_g is measured in mechano-Henries per seconds (L_g/dt) which corresponds to a negative shift in space (something like a displacement of the space metric backwards) per second $(-\Delta x/dt)$. Likewise, the ε_g is measured in mechano-Farads per seconds (C_g/dt) which corresponds to a shift forward in space (like a displacement of the space metric forward) per second $(+\Delta x/dt)$. Furthermore, the involvement of a shift in space $\pm\Delta x$ within the terms ε_g and μ_g indicates that **the swapping of the parameters of space and time in (5-44) gets this interaction to be dependent to the involvement of the Weinberg angle**, as the corresponding phase shift in space concerns this angle.

It is of importance that since either one of the shifts $-\Delta x$ and $+\Delta x$ is in the opposite direction with respect to the other, they concern opposite sign of energies, therefore they behave as either one of them is imaginary with respect to the other. This is why only one of the two, in this case μ_g, is reciprocated in the equation (5-44), since it behaves as an imaginary entity that interacts with the imaginary part of a particle's wave function, and projects to the world as a real entity due to the relation $i^2=-1$, so $i=-1/i$.

Equation (5-44) may allow to figure the velocity of propagation of a GMW. For the propagation to take place without impedance, the values of L_g and C_g should be in balance to each other, so neither of them applies a net mechano-inductive or mechano-capacitive impedance. (This is analogous to the case of propagation of an EMW). By substituting $\varepsilon_g=+\Delta x/dt$ and $\mu_g=-\Delta x/dt$ to (5-44), we get:

$(-dt/\Delta x)\, dT/dt = (+\Delta x/dt)\, dg/dx$, which reduces to:

$$dT/\Delta x = \cancel{(+\Delta x/dx)}\, dg/dt \qquad\qquad (5\text{-}47)$$

Here, if we loosely accept canceling out the term $(+\Delta x/dx)$, which we have already struck through on the basis of the similarity in the units of Δx and dx both in "meters", then (5-44) gets to look relatively similar to (5-42). This resemblance exists because the units of the mechano-permittivity and mechano-permeability are such, that their involvement in (5-44) resembles a swapping of the parameters of space and time in the denominators of (5-44). However, we should not overlook the fact that the denominator in the left side concerns a phase shift IN space Δx, instead of WITH RESPECT to space dx which is the case for (5-42), so they account for different attributes. Subject to such inaccuracy, we get (5-47) look very much similar to (5-42). Or, likewise, (5-42) gets to look effectively like a limiting case of (5-44).

This special case holds if C_g is in balance with L_g, so that the value of $(+\Delta x)$ of the former has equivalent effective magnitude with the $(-\Delta x)$ of the latter, so they may both be loosely represented by the simplified notation dx, and in this case (5-44) resembles (5-42). Note that if that condition of equality is not met, then the difference in value may reflect on a displacement (tunneling) of negative momentum, as we shall consider in section 6. As we shall figure then, that is possible since a shift (L_g or C_g) is typically represented by a complex term (including a real as an imaginary part) where the imaginary part may account for the displacement of momentum in a similar sense to as charge is displaced between the plates of a capacitor (to be addressed ahead).

Regarding the propagation of GMWs, just as it was previously considered for the case of the EMWs, the propagation does not exactly concern the transmission of plain alternating gravitational and gravitomagnetic fields. The engagement of ε_g and μ_g get the actual fields involved to correspond to a **mechano-capacitive displacement gravitational** field $g_d=\varepsilon_g g$ and a **mechano-inductive displacement gravitomagnetic** field $T_d=(1/\mu_g)T$. In fact, if one looks to the units involved, it arises that what we may (illusively) consider an alternation of T as per (5-48), is actually an alternation of g_d^+ as per (5-49):

$$T = c\, g = (dx/dt)\, g \qquad\qquad\qquad (5\text{-}48)$$

$$g_d^+ = \varepsilon_g\, g = (\Delta x/dt)\, g \qquad\qquad\qquad (5\text{-}49)$$

And likewise, what we may illusively consider to be an alternation of g as per (5-50), is actually an alternation of T_d^- as per (5-51) below. In fact, notice that (5-51) involves a phase shift backwards in the space metric (mind the minus sign in $-\Delta x$), and since this shift backward refers to an imaginary entity, the mechano-permeability becomes involved in the form of a reciprocal.

$$g = (1/c)\, T = (dt/dx)\, T \qquad\qquad\qquad (5\text{-}50)$$

$$T_d^- = (1/\mu_g)\, T = (dt/-\Delta x)\, T \qquad\qquad\qquad (5\text{-}51)$$

It therefore seems to be more accurate to describe the propagation of a travelling GMW in a form like (5-52), rather expressed in the form (5-53):

$$dT_d^- /dt = dg_d^+ /dx \quad \text{or rather} \tag{5-52}$$

$$dT_d^- /(1/dx) = dg_d^+ /(1/dt) \tag{5-53}$$

In this case, the involvement of mechano-capacitance and mechano-inductance within the oscillating g_d^+ and T_d^-, translates in that the oscillation of the displacement gravitational and gravitomagnetic field versions goes along with a concurrent oscillation of a phase shift (a displacement) in space. In such case, **GMWs may propagate through a repetitive step-by-step shift (displacement) in the space metric. That is, along with every alternation between the g_d^+ and T_d^- fields, there comes a positive shift and then a negative shift in space, both pointing in the same direction** (the direction of GMW propagation). In addition, that the mechano-permeability term contained within the left side of in (5-53) is reciprocated, seems to reflect a geometric inversion which provides to the field "lines" of T_d^- a radial-like character. All the above, therefore, appear to provide a differentiated understanding with respect to what is currently thought of about the way of propagation of GMWs.

The repetitive shift in space is responsible for getting the GMWs propagate sideways to the direction of the effective acceleration or matter which emitted it (either plain oscillatory, or circular). In this sense, the emission of the wave is actually an effect of reaction, and as we have considered already, **a reaction always develops at an orthogonal direction with respect to the action that causes it**, for reasons of the balancing of energy (this concerns the orthogonality between imaginary-to-real axes). In that specific direction, each progressive mechano-capacitive shift in space $+\Delta x/dt$ and each mechano-inductive shift in space $-\Delta x/dt$ are in balance to each other, so the wave may propagate without impedance. And in relation to that, **no force (in the form of "displacement momentum") rides the wave in this specific direction**, and its fields are considered radiative. (Note that in the contrary, a non-radiative case will be considered in section 6 to concern the effect of barrier penetration. While, a somewhat similar condition also becomes involved in the weak interaction as we shall consider in section 9).

Furthermore, the **velocity of propagation of GMWs** may be figured if we re-arrange the terms of relation (5-44), so it becomes $dT/dt = (\varepsilon_g \mu_g) \, dg/dx$. Then, in symmetry to what was the case with EMWs, the velocity of GMWs should arise from (5-54), where the balancing between L_g and C_g in Space should apply in same way to the balancing between L and C in time for the EMWs, in which case the minimum (zero) impedance applies at the speed of light.

$$\upsilon_{(GMW)} = \left(\varepsilon_g\,\mu_g\right)^{\frac{1}{2}} \tag{5-54}$$

If the above condition of balance is not met, then the corresponding fields obtain a coordinate toward the direction of the matter's motion, where that coordinate is non-radiative, and different equations apply for it, which may involve the displacement of negative momentum (when an imaginary field version is involved). However, since an imaginary field version concerns negative energy, it damps out quickly and becomes evanescent in the real world (just like it holds with displacement fields in barrier penetration, which we shall consider in section 6).

On a point of attention, the relations that we have provided above concern a classical physics approach, which differs to the quantum mechanical approach, in that the interactions develop with respect to **unitary** space and time (fixed values of the metrics of space and time which prevail at our cosmic neighborhood), referring to specific fastness of passing time and space stretchiness applicable in equations' denominators dt and dx. This is what gets GMWs interact in the form of classical waves. While, when the corresponding fields interact at the atomic level, they interact in **non-unitary** space and time and that brings up a quantized behavior. That concerns an inversed type of interaction involving oscillations in space and time, as we shall figure in sections 7 to 9, where we shall also address the case of the "graviton".

Point 4: **Classical vs. Intrinsic gravitomagnetic field**

As per the illustration of figure 5-4, even though electrons appear to be matter monopoles (of conventional positive matter), they should be considered having an imaginary hidden part too (concerning negative matter) which relates to the imaginary portion of the particle's wave nature. The conventional matter part links to the particle's gravitational field which has radial-like field lines, of which we perceive only the real part extending from the particle toward infinity, and we do not perceive the remaining part which is imaginary to us. The negative matter part, instead, becomes involved in effects of reaction (concerning negative energy), through a field version which has imaginary character and loop-like field lines (due to the shift involved in mechano-permittivity and mechano-permeability terms). That field engages in the generation of vortex flow (in symmetry to the generation of current in a metal ring by the loop-like electromotive potential in electronics).

Just like the above holds for the gravitational field, a similar differentiation applies for the gravitomagnetic field as well, which also comes in two versions. A conventional "classical" ordinary gravitomagnetic T which has loop-like field lines, and the opposite-energy gravitomagnetic $\bar{T}$ field which interacts through an inverse geometry (due to the involved shifts), whose field lines are radial-like. Basic characteristics of these two field versions have as follows:

Classical Gravitomagnetic field T
The classical (ordinary) gravitomagnetic field T is associated to the deflective motion of ordinary matter, and its field lines are direction degenerate, as shown in figure 5-12 left part (gray coaxial lines). As mentioned earlier, when the density of the field lines is not evenly balanced around the flow (due to external reasons and influences), the difference in the density of field lines causes sideways deflection of moving (flowing) matter, like in the lift of the wing, or in the sideways propagation of tornadoes.

Fluid flow is often studied while it flows along a straight line (like in figure 5-12 left part), however effects at larger scale often describe a more complete picture of fluid flow over a whole loop. That favors a better understanding of the complete gravitomagnetic interaction, and helps in preparing formulas to describe it. (This holds in a similar manner to as the magnetic interaction is more thoroughly studied in the case of a current in a complete circuit loop, that in the current in a local straight-line segment). Actually, at the very large scale, like at celestial or galaxy scale, we might even have the motion of solids (in that case, celestial bodies) follow the rules of fluid flow (which is not exhibited in smaller solids, since particles there don't have adequate motional freedom to displace in respond to local changes in pressure).

But then, one may wonder if it is really feasible for the gravitomagnetic field to apply at the celestial scale. The principle behind the gravitomagnetic interaction involves a shifting in the parameter of space, and space exists even in conditions of vacuum, so in this respect there should not be a restriction for the engagement of the corresponding potential in the open vacuum. There is however another practical limitation, because T is a negative energy (imaginary) field, therefore it applies only marginally (at close range) in the classical world, which is why its action is met mostly in surface effects, while beyond that close range it quickly becomes evanescent. Notice that in cases like the one of the airplane wing, the presence of a very large number of air molecules at close range to each other lets the gravitomagnetic interaction (the Bernoulli pressure) disperse over larger distance because each air molecule affects its neighboring molecules, therefore each molecule serves as a stepping stone for the total effect to be transmissive over a larger distance away from a surface. But in the absence of a fluid, the effect should rather not have long range over vacant space. In that sense, at very large scale (like at celestial scale), the extreme scarcity of matter in-between celestial objects should ordinarily make any effects of the corresponding force non-existing.

This limitation in terms of the range of action of this field could however be overcome when an inverse-energy version of the effect takes place, in which case the "opposite of the negative" energy lets the effect bring outcome in the real-world at large range (and scale), as we shall consider right next.

Opposite energy (intrinsic) Gravitomagnetic field $\bar{T}$
The opposite-energy gravitomagnetic field $\bar{T}$ differs from the classical T in a similar way to as $\bar{B}$ was explained to differ from B. We shall here consider this field in two different types of interactions. One concerns interaction with negative energy particles like neutrinos in **unitary spacetime** (in specific stretchiness of space and fastness of time which prevail at large scale and are described by classical-like conditions). Another type concerns interaction with the imaginary part of particles in **non-unitary spacetime** (concerning elementary particle's quantum nature). We shall start with the latter.

While this field is of negative energy (imaginary to the classical world), it may engage in interaction with the imaginary part of the particle's wave nature and therefore project to the real world as a real field due to the relation $i^2 = -1$. This projection to the real world is made possible due to the existence of the Space-to-Time shift shown in figure 1-1 (the Weinberg angle), and is subject to a Sine coordinate. Since the Weinberg angle corresponds to 30°, the projection to the real world is subject to a factor of $\text{Sin}(30°) = \frac{1}{2}$. This dependence is what gets electrons exhibit a **gravitomagnetic moment of** $\frac{1}{2}$ when they interact at the atomic level as fermions.

According to theory to follow, stationary electronic states in atoms do not involve conventional motion of the electrons, instead the negative gravitomagnetic field may be understood to arise from the particular wave-like behavior of atomic electrons which is completely different to conventional motion of matter. It concerns an oscillation in the metrics of time and space themselves, instead of an oscillation "with respect to" space or time. As we shall explain in sections 7 to 9, this is the reason why electrons may occupy electronic states within atoms without exhibiting conventional angular momentum.

Furthermore, the negative energy of the negative gravitomagnetic field gets it interact in the real world via an inverse geometry because of the phase shifts within the mechano-permittivity and mechano-permeability terms, so its field lines are radial-like, instead of loop-like as that of the conventional gravitomagnetic field. In this case, we are able to perceive only the part of its field lines which extend from the particle in the direction toward outer space, and we do not perceive the other part which is imaginary to us. This characteristic also gets this field apply as the field of a distinct gravitomagnetic pole (looking much like the magnetic field of a bar magnet's pole, which exhibits radial-like field lines, and not like the magnetic field of an electric coil which exhibits loop-like field lines).

The opposite sign of energy and the corresponding reactive nature of the gravitomagnetic field in this case, gets it apply in the form of a **"gravitomagneto-motive"** potential. And that potential seems to relate to the Van der Waals forces, which concern a mechanical spin-polarization interaction of nearby particles. These behave much like tiny bar magnets but their forces point in opposite direction, attracting particles of same spin gravitomagnetic moment of ½, while repelling away ones of opposite spin. They apply as weak attractive bonds (weaker than covalent or ionic) and quickly vanish at longer distances between molecules (which is what should be expected, since beyond that range the corresponding potential should become evanescent due to its imaginary nature with respect to the real world). The gravitomagneto-motive potential seems to apply in the opposite direction to the magneto-motive one which was stated to be responsible for the Pauli exclusion principle where that potential was repulsive for particles of same spin, instead of attractive, and seemed to be attributed to the intrinsic magnetic field as explained earlier. Examples of such interaction may include bonds in organic compounds (typically involving Hydrogen) which relate to the powering of living conditions, with more to be said on this topic in section 6).

While the above are concerning the interaction of this field with fermions in non-unitary spacetime, another case concerns the involvement of the negative gravitomagnetic field in interaction with particles of negative energy in conditions of unitary spacetime (classical-like). This concern interaction with neutrinos and antineutrinos (in cosmic regions that constitute their habitat), as well as the interaction with quarks. In section 10 we shall particularly describe that quarks correspond to lepton-like particles which incorporate a capacitive-like or an inductive-like shift in their parameters, which makes them behave partially as electrons or positrons, and partially as neutrinos or antineutrinos. The interaction with the negative energy portion of quarks, as well as with neutrinos (both of which involve opposite energy), allows for the action of the imaginary gravitomagnetic field to extend over long range (and scale), even at galaxy level. The generated difference in space curvature that this potential corresponds to, brings a radial-like effect resembling a form of frame-dragging.

The projection of such an imaginary-born interaction to the real world still needs to involve an inverse geometry, and in relation to that, the strength of the negative gravitomagnetic action force should change with respect to reciprocal space ($1/dx$) instead of with respect to space dx. That gets the corresponding potential to apply with respect to the reciprocal of the radius. As we shall be considering a little later in this section, this kind of interaction seems to account for the "missing force" required for explaining the "flat-curve" rotational motion of galaxies, which raised the need to search for the supposedly missing hypothetical **dark matter**.

Point 5: **How the gravitomagnetic field arises**

Just like its magnetic counterpart, the conventional gravitomagnetic field T is considered to arise either via the motion of matter (which may be a steady state effect), or via the alternation of gravitational field/flux (a transitional effect). The latter was considered earlier through the equations (5-39) and (5-40) and will be further elaborated in section 6, while we shall here emphasize on the former, which concerns two types of cases:
- The formation of the classical gravitomagnetic field around the flow of conventional matter, which will be explained to arise through a process which is symmetric to the one described by Ampere's law (5-3). This involves the motion of electrons' real portion, in association to the particle nature of electrons.
- The formation of the negative (intrinsic) gravitomagnetic field, which has orthogonal properties with respect to the classical one, and its field lines are radial-like. It turns out that this variant ascribes to the electrons' oscillation in non-unitary spacetime (which does not concern conventional motion, it concerns an alternation in the metrics of space and time), so it relates to electrons' wave nature.
More particularly:

Gravitomagnetic field of particle nature
We have so far described how the T field develops around the flow of matter on a straight-line segment (as in figure 5-12 left part), but we could get a more complete picture if we considered the effect taking place over a whole loop of fluid flow. To approximate that, lets us refer to a particular paradigm from electromagnetism. Consider a metal ring. If a magnet is approached to it, this changes the magnetic flux Φ_B through it, and while this is happening, a current develops through the ring according to Lenz's law $\oint E \cdot d\ell = -d\Phi_B/dt$ (where that current creates a magnetic field which tends to cancel out the imposed

change in magnetic flux). If the opposite holds, and a *current* is imposed in the metal ring (instead of a change in *magnetic flux*), the changing current would cause a change in the magnetic flux through the ring. In either case, nature tends to react a change, and retain the flow of current unchanged. This may look like a law of conservation of current (applicable for ideal conditions of current flow), however, as previously mentioned, nature rather tends to conserve the magnetic flux $\Phi_B = LI$ (and not the current alone).

It turns out that a symmetric principle applies in the case of mechanics as well. As known, nature tends to conserve momentum (and angular momentum too), however we shall postulate that nature tends to conserve a quantity that we shall call gravitomagnetic flux Φ_T, and not momentum alone. The gravitomagnetic flux may be represented as follows, where L_g is the mechano-inductance and p corresponds to the momentum of the moving matter/flow.

$$\Phi_T = L_g\, p \qquad\qquad\qquad (5\text{-}55)$$

In a case of a loop-like fluid flow, the gravitomagnetic flux throughout the loop also tends to remain constant, and this is accomplished if the circulating matter does not change angular velocity. This is illustrated schematically in figure 5-15, where the steady flow of matter in fluid form (thick black arrow) has associated with it a mechano-capacitive potential around it (radial-like arrows pointing inward) which concerns the Bernoulli pressure, and this creates equipotential surfaces which correspond to the gravitomagnetic field T (grey coaxial circles around the flow).

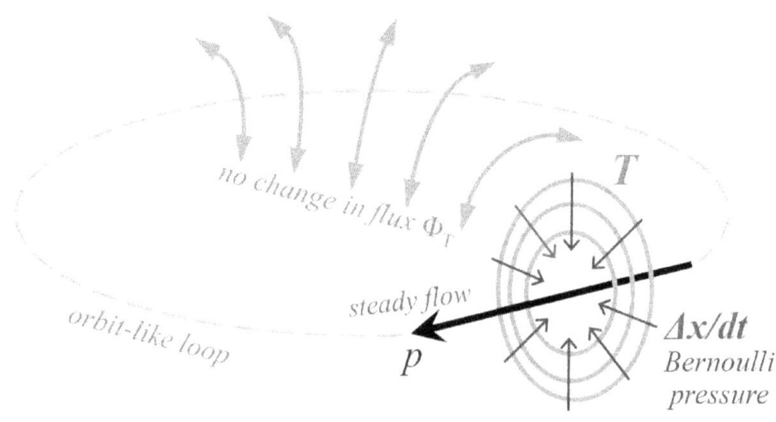

<u>Figure 5-15</u>: Fluid flow creates a gravitomagnetic field T around it. The gravitomagnetic flux Φ_T through the loop tends to stay conserved.

Notice that the field lines of T do not have an arrow to indicate a particular direction of turn, this is because they are direction degenerate, as either direction of turn works equally fine (since only positive sign of matter prevails in the classical physics world).

Just like the case was in electronics, a change in the radial-like mechano-capacitive differential is counterbalanced by a mechano-inductive differential which may deploy loop-like over the orbit-like loop. As matter in fluid form moves (flows) in an orbit-like path, the gravitomagnetic field of this whole loop should look like the magnetic field lines of a metal ring or coil (not shown in this figure). For reasons of conservation, the momentum of the flow throughout the whole loop tends to remain unchanged, as any change is reacted through the mechano-inductance L_g, (which applies much like a loop-like electromotive potential), hence the gravitomagnetic flux Φ_T through the area of the loop tends to remain conserved.

In accordance to the above, an Ampere-like law may arise via the following pattern, and describe how the gravitomagnetic field forms around a steady momentum flow p, over the loop in time dt. The loop in time coincides to the orbit-like loop of the momentum flow, however this notion will become more transparent later, when we shall explain vortex flow, in the next section.

$$\Phi_T = L_g\, p$$
$$T\, dt^2 = (-\Delta x)\, p$$
$$dt/(-\Delta x)\, T\, dt = p$$
$$1/\mu_g \oint T \cdot d\ell_t = p \tag{5-56}$$

Notice a particular symmetry with Ampere's law $1/\mu \oint B \cdot d\ell = I$ where one can -once again- figure that the mechanical and the electrical cases differ to each other by a swapping of the parameters of space dx and time dt. That is:

- In the mechanical case, the right side of (5-56) concerns the momentum p, which corresponds to $p = h/\lambda$, where h is the energy concerning the Planck constant, and λ concerns the particle matter wave's wavelength (which could be represented as dx). In this sense, the right side of this relation describes a **quantized entity concerning a particle's matter (h), per space (dx)**.
- In the electric case (Ampere's law), the right side concerns the current $I = Q/dt$, which describes a **quantized quantity concerning a particle's charge (which for an electron corresponds to $Q = 1e$), per time (dt)**.

The above relation (5-56) concerns a *steady* loop-like fluid flow of matter (thus the angular momentum is being conserved). If angular momentum is forced to change, that needs to be counterbalanced through an effect of reaction. The corresponding reaction would need to take place in an orthogonal direction (since effects of reaction are stated to have imaginary character with respect to the action that they impede, same as an imaginary axis is orthogonal to a real axis). And also, it should have a temporal character, lasting only for as long as the change takes place. Through the illustration of figure 5-15 one may realize that if a change in the magnitude of fluid flow takes place (in the direction of the thick black arrow), that would need to go along with a change in gravitomagnetic flux Φ_T through the orbit-like loop. In such case nature tends to create a "gravitomotive" potential to oppose the imposed change in flux. This reaction concerns a relation of the general form $\oint g \cdot d\ell = -d\Phi_T/dt$, which we shall comment upon in the next section. As we shall see, that is analogous to Lenz's law of electromagnetism, and the corresponding reaction concerns processes of vortex-like flow.

Gravitomagnetic field of wave nature:
The intrinsic gravitomagnetic field has certain orthogonal properties with respect to the classical one, like that its field lines are radial-like as described earlier. It turns out that this field version relates to electrons' wave nature which concerns an alternation IN the metrics of time and space, instead of an alternation PER time or space. That alternation of metrics is actually that makes it possible for the wave form of the particle to settle in an electronic state without exhibiting conventional angular momentum. The term "electronic" state seems like referring to the charge-related properties of electrons, but we shall focus on the "mechanical" aspect of that state, meaning the matter-wave part of wave nature. While the particulars of this alternation shall be considered mainly in section 7 in relation to the Schrödinger equation, we may here provide a first approximation.

Since we are going to address wave-like properties, we need a wave equation to use as a base of reference, and we could attempt to make an **inversed reading** of it so as to account for the oscillation of space and/or time. In particular, since we focus on the particle's matter, we shall refer to aforementioned mechanical relation which is symmetric to Maxwell' s fourth law which we presented above in (5-40), and will copy it here for reasons of convenience, along with its simplified expression (5-44) which we also used earlier:

$$(1/\mu_g)\oint T \cdot d\ell_t = (\varepsilon_g)\, d\Phi_g/dx \qquad\qquad \text{(5-40) or (5-57)}$$

$$-(1/\mu_g)\ \boldsymbol{dT/dt} = (\varepsilon_g)\ dg/dx \qquad\qquad\qquad \text{(5-44) or (5-58)}$$

Let us focus on (5-58). Considering that the mechano-permeability μ_g corresponds to mechano-Henries per seconds (that is, L_g/dt) and the mechano-permittivity ε_g corresponds to mechano-Farads per seconds (that is, C_g/dt), we could imagine a differentiated reading of this equation, where the right part may describe an oscillation in mechano-capacitance C_g (instead of an oscillation of the gravitational field g), and the left part of the equation may describe an oscillation in the reciprocal of mechano-inductance L_g (instead of an oscillation of the gravitomagnetic field T). That could loosely refer to the same relation now being red in the form of

$$T\ [-d(1/\mu_g)/dt] = g\ [d(\varepsilon_g)/dx] \qquad\qquad\qquad \text{(5-59)}$$

Such a reading, may concurrently associate to an inversion in behavior. That is, instead of reading (5-58) as having oscillatory values of T and g and fixed values of mechano-inductance L_g and mechano-capacitance C_g, (where that condition ascribes to **bosons**) we come up with an opposite kind of condition; an oscillation of the values of L_g (that is, and oscillation in the shift in negative space) and C_g (that is, an oscillation in the shift in positive space) along with fixed values of gravitomagnetic T (associating to the particle's intrinsic gravitomagnetic moment which has a fixed value) and a fixed gravitational field g (where this condition ascribes to **fermions**).

Even though such a modified reading of the above equations is oversimplified, it does provide an indicative approximation of what holds for atomic electrons in terms of an oscillation in time and/or in space. It also paves the way for understanding the procedure that concern the formation of the intrinsic gravitomagnetic field. It seems that the wave formation representing the electron matter around an atomic nucleus concerns a waviness in the fastness of imaginary time. In this way, instead of an electron moving with respect to time, we have an oscillation OF time itself, with respect to the electron's formation. In this way we get the familiar electron lobe-like representations without actual motion of matter. In counterbalance to this, we have the formation of a particle's fixed gravitational field. This identifies to **a waviness of time as that projects (deploys) over space, constituting the particle's gravitational field**. A similar type of waviness in space (a mechano-inductive effect) takes place in parallel and accounts for the particle's intrinsic mechanical spin and the associated gravitomagnetic field produced. This again, takes place without actual motion of matter, but with the alternation in the stretchiness of space causing a space dragging effect which concerns the intrinsic gravitomagnetic field. Here, this radial-like field accounts for a centripetal-like balancing action that keeps the particle's matter-wave in a loop-like formation. (This is in symmetry to what was explained to hold with respect to the particle's charge and the radial-like intrinsic magnetic field which keeps electron's electronic state balanced in a loop-like formation). In that sense, equation (5-57) may be interpreted in a new complementary way as follows:

- In the left side of the equation, the integral should NOT refer to the integration of the field along the whole path of the orbit-like loop (shown in figure 5-14). Instead, it should rather refer to a path over the RADIAL-like direction (looping through the very distant universe), addressing the intrinsic gravitomagnetic field which may act as a centripetal-like force that keeps the particle's matter balanced in a loop-like formation, and associates to its fermionic behavior.
- In the right side of the equation, the flux should rather concern a LOOP-like imaginary version of the particle's gravitational field. In the present case (which concerns non-unitary spacetime, and the alternating entity is the mechano-capacitance), this side of the equation should rather associate to the particle's fixed intrinsic spin. (Note that in another condition which concerns unitary spacetime, the same type of field seems to associate to the generation of vortex flow, which we shall consider in the next section).

Due to the orthogonal character of the intrinsic gravitomagnetic field (with respect to the classical gravitomagnetic field), it plays its own contributing part to matter's gravitomagnetism, independently from (but in conjunction with) the classical gravitomagnetic field created by conventional motion of matter.

Point 6: Reconsidering Dark Matter

The concept of dark matter arose following calculations that the observed gravitational mass in the universe was not sufficient to account for keeping galaxies together. Because of that, it was hypothesized that there should be more matter present. And since that matter could not be detected, as it was not luminous, hence the term "dark" arose.

Looking into the issue more closely, according to Kepler's Second Law which accounts for the orbital motion of the planets in our solar system, it would be expected that the larger the distance from the center of a galaxy, the rotational velocities of stars should decrease, similar to as it holds for the planets of our Solar System (figure 5-16, left part). This is based on the law of conservation of angular momentum L (not to confuse angular momentum with inductance which uses the same letter-symbol). According to this law it holds $L=rmv$

where r is the radius of rotation and mv is the linear momentum. So, if the radius is reduced by half, the velocity should double in order for angular momentum to be conserved. This is what holds for the planetary model, and it is the same principle as for an ice skater that pulls her hands in and spins faster. In the contrary, at much larger scale the observations of galaxy rotations reveal that the rotation curve in spiral galaxies remains flat as distance from the center increases [15], [16], which means that the rotational velocity of stars is not getting lower the further away from the galactic center (figure 5-16, right part).

According to calculations, this discrepancy could be resolved if the gravitational force experienced by a star in the outer regions of a galaxy was proportional to the *square* of its centripetal acceleration (as opposed to the centripetal acceleration itself, like in Newton's second law), or alternatively, **if the gravitational force came to vary inversely with radius** (r) as opposed to the inverse square of the radius (r^2) as per Newton's law of gravity [17].

According to that, the galaxy rotation could be possible to explain without the need for dark matter, if some other attractive force of proportional strength would apply on top of the gravitational, where that new force would vary with respect to "**reciprocal**" radius ($1/r$), meaning that it would be stronger the further away from the center of the galaxy. In such way the total attraction force, including both the gravitational and that other force, would vary (approximately) with respect to $(r^2) \times (1/r) = r$. To explore such possibility, we need to figure: (i) what could be the nature and origin of such other force and if that may concern negative energy, and (ii) why it seems to apply at the galaxy scale but not at the solar-system scale.

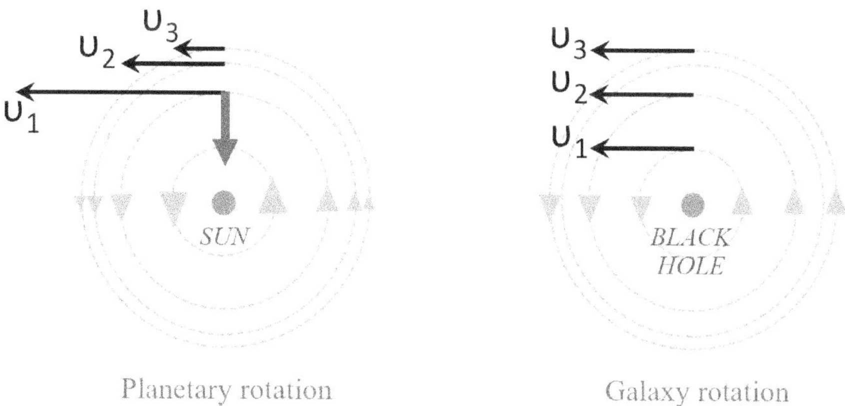

Figure 5-16: Rotation speed of planets around a star (left) versus rotation speed of stars of a spiral galaxy (right).

One thought to explore is that, since the waviness in the *time* metric (described through figure 1-1) is stated to concern gravity, could the waviness in the curvature of positive *space* possibly account for such an additional force. If that could be the case, then such an effect of space would also need to provide an answer to the question (ii) of why would that force act at galaxy scale, but not at the smaller-sized solar system scale, where Kepler's law applies. Starting from the later point, a candidate explanation could concern the phase shift between the time and space curvatures of figure 1-1. In particular:

- At a relatively small celestial scale like that of our planetary system, the radius of revolution of planets around our sun is far smaller than the spatial extend of the phase shift between the space and time curvatures depicted in figure 1-1. (This is more obvious when this shift is measured in terms of distance, and not in terms of degrees of angle which refer to the Weinberg angle). Therefore, at the scale of our planetary system, there is almost no influence of such a phase shift centered on our sun, so that practically leaves only the space curvature of our mother galaxy's black hole to be in effect. Because of that, the space curvature in our celestial neighborhood is relatively flat, and leaves it only for gravity to be the main force that keeps our solar system together. Therefore, Kepler's second law applies, where the force varies with respect to the square of the radius (as shown at the left side of figure 5-16).
- At a galaxy scale instead, the radius of a galaxy is far larger than the spatial extent of the Weinberg shift, and space goes through a node at point b of figure 1-1. In this case, the curvature of positive space outside that nodal surface may acquire significance, and if that may relate to some appropriate potential, then the combined attraction at galaxy scale (due to gravity and due to such potential) could get Kepler's law be no more valid, as a different balancing applies, according to which the rotational velocity of stars does not decrease the further away from the galactic center.

To further support the above supposition, we would need to specify what type of curvature effect could constitute the origin and nature of the corresponding potential. If that were a *space*-curvature-related potential and force which applies in symmetry to the gravitational *time*-curvature-related force, then its action (from the applicable range onwards) should rather vary inversely with the square of the radius (r^2). But that option does not match the existing observations revealing that galaxies' rotation curve remains flat as the distance from the center increases. This calls to look for some other possibility for this potential's origin. In that case, another possibility to consider could concern a gravitomagnetic-related force. For that case, the same reasoning (of the phase shift) would still hold in explaining why that potential and force applies at galaxy scale but not at solar system scale.

As considered above in relation to figure 5-12 left part, the gravitomagnetic field lines correspond to equipotential lines of a radial-like potential, which comprises of a differential of a *space shift* $\Delta x/dt$. (That comes in symmetry to the magnetic field lines corresponding to equipotential lines of a radial-like potential which comprises of a differential of a *time shift* $\Delta t/dx$ as shown in figure 5-12, right part). A significant difference between the magnetic and gravitomagnetic field cases, however, is that the former has the characteristics of a *real* field so it may apply at long range, while the gravitomagnetic field T has the characteristics of an *imaginary* (negative energy) field, so it quickly becomes evanescent in the real world, thus it has short range of action as explained earlier.

But let us suppose that the gravitomagnetic interaction could also take place in conditions of opposite energy, that is, apply on particles' imaginary portion (associating to negative matter, like the one that neutrinos are made of). That could have the interaction of two imaginary entities to bring a real result, in which case such interaction: (i) could have long range of action, and (ii) would project to the real world with respect to reciprocal conditions (per $1/dx$ instead of per dx). A way that such condition could take form, is if the negative (imaginary) gravitomagnetic field could interact with motional flow of negative form of matter. We therefore need to explore if such an inverse type of interaction could apply in some way between imaginary entities, and generate a capacitive-like attractive potential that would account for the missing attractive force (which raises the need to hypothesize the existence of **dark matter** in the first place).

For an opposite-energy gravitomagnetic field to be generated at a galaxy scale, it should relate to a flow of particles of negative matter, so that flow would create around it a radial-like space dragging effect $-\Delta x/dt$, similar to as it holds with the Bernoulli pressure around a fluid flow, now applicable at the large scale (figure 5-17, left part). At first glance, there seems to be no observational proof of such flow to be taking place at a galaxy scale, to make this an actual scenario. (And, neither an alternative -rather remote- hypothesis, of the motion of a whole galaxy through vacant space could make for an equivalent solid scenario).

What seems to make a realistic case, concerns the flow of particles toward opposing directions, like at both sides of an accretion disk, in a mirror-like way (figure 5-17, right part). Such an alternative could generate a radial-like space potential of $\Delta x/dt$. Actually, the flow of particles toward opposite directions does not create any issue, since the gravitomagnetic field is direction-degenerate. In fact, such kind of mirrored flow of particles and energy complies to the observed process of astrophysical jets developing at both sides of accretion disks.

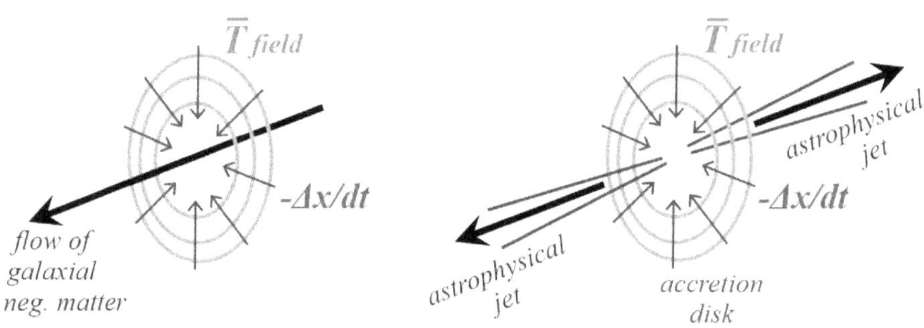

Figure 5-17: Negative gravitomagnetic interaction scenarios at galaxy scale, concerning straight flow (left), vs. mirrored astrophysical jet flow (right)

In such case, the next thing that we would need to figure, is how the interaction could involve negative energy, and negative sign of matter. According to observations, astrophysical jets correspond to outflows of ionized matter emitted along the axis of rotation close to the speed of light. These are considered likely to arise from dynamic interactions within accretion disks where energy is extracted through relativistic effects like frame dragging. Such active processes are commonly connected with objects such as central black holes of active galaxies, quasars, supermassive black holes, neutron stars, and pulsars [18],[19]. In accordance to what was presented in section 4, the neighborhood of such celestial objects corresponds to regions like b - b′ or d - d′ where the product of the Time and Space metrics is negative (see figure 4-2), so negative energy effects should be considered ordinary.

According to existing theory, these particular areas seem to relate to "ergospheres", and a space dragging process takes place there [20], which has the implication of the existence of negative energies within the ergospheres. And in relation to that, an object within the ergosphere cannot appear stationary with respect to an outside observer at a great distance unless that object were to move faster than the speed of light with respect to the local spacetime. This, in fact, seems to come in certain compliance to the present theory, since the processes in these regions should involve matter of negative sign, and as explained in section 4, that cosmic region should constitute stable habitat for particles of negative matter. Such matter of negative sign should actively engage in the creation of the astrophysical jets, and constitute a primary constituent of their essence. That assumption would suggest that these jets should comprise of contents as follows:

- In terms of **leptons**, the jets should eject particles like neutrinos, which are of negative energy. (Notice that the region adjacent to the black hole is of negative space metric, so antineutrinos and neutrinos should constitute "real" matter within that region).
- In terms of **baryons**, the jets seem to be ejecting matter of negative energy (in parallel to matter of positive energy), and this is possible due to a very specific property of baryons' constituent particles, the quarks. As we shall introduce later, quarks' nature extends in both positive and negative energies (so a portion concerns negative energy). That will be considered mainly in section 10, but we may here provide an overview about these particles' unusual nature, as follows.

As we have described so far, leptons like in particular the electrons, have a real part (comprising matter) and an imaginary part (which remains hidden as it does not take real form), and the imaginary part may become involved in effects of reaction (to be considered in section 6) thanks to the phase shift between and space and time metrics shown in figure 1-1. Likewise, in section 10 we shall figure that such a phase shift may also apply *internally* to a particle as it concerns the oscillatory parameters of the time and the space in its wave function. This allows (subject to conditions) for the creation of a different type of particles which are lepton-like, but their real portion is shifted partially toward the imaginary portion, and concurrently their imaginary portion is shifted partially toward the real portion. This also affects the value of charge and electric field that these particles exhibit (and similarly applies for their other fields). This type of particles correspond to quarks, which are much like "leptons which are subject to an internal phase-shift", and this feat gets them to consist partially of positive energy and partially of negative energy. In this case, their negative energy portion seems to engage in a negative gravitomagnetic field interaction, which we are looking for.

Under a similar way of illustration as in figure 5-5 left part we depicted a conceptual "side-view" of an electron (with the upper one-half concerning the real/visible part of its electric field and the lower one-half concerning its imaginary/non-visible part), now a corresponding "side-view" of quarks is schematically illustrated in figure 5-18. Remember that the field ascribed to a single particle concerns the part *at the left* of the vertical dashed line. In the figure's left side, one may notice that the "negative electric field" (black arrows) have been phase shifted clockwise (forward) by 60°, leaving only 1/3 of the real electric field active (the black arrows ascribing to a value of charge $Q = -1/3$ e). Likewise, at the right side of the figure one may notice that the "positive electric field" arrows have been phase shifted clockwise (backwards) by -30°, leaving only 2/3 of the real field active (black arrows ascribing to a value of charge $Q = +2/3$ e). The rest of the arrows (grey arrows at the left of the vertical dashed lines) correspond to anticharge, which is imaginary. And just like it holds for the particle's charge, it similarly applies for the particle's matter as well, where a part of a quark's essence concerns positive matter and the other part concerns negative matter (which is different to antimatter).

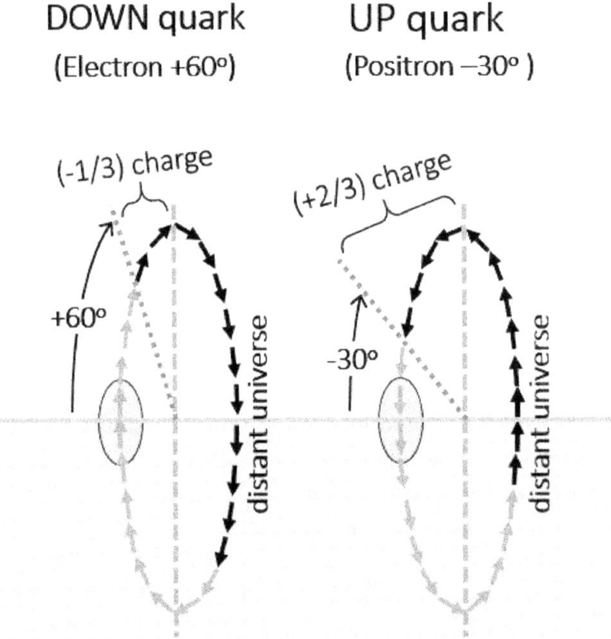

Figure 5-18: Schematic depiction that a phase shift of a particle's fields by 30° or 60° transforms a lepton into a quark

Along these lines, figure 5-18 allows to schematically exhibit that a portion of quarks "lives" in negative energy. And this portion should therefore be open to interactions with fields of negative energy (described by imaginary terms). Thus, the two interacting imaginary entities (the negative part of the particle and the field of negative energy) may bring up a negative real outcome, as per the relation $i^2 = -1$. And in relation to that, the motion of the astrophysical jets should generate around them a radial-like, negative space-shifted potential of $(-\Delta x/dt)$ extending beyond the region of the so-called ergosphere, which should act as a potential which tends to displace the imaginary part of particles inwards (in the direction toward the central black hole). The process concerns a negative space dragging effect, which applies in a similar way to as the Bernoulli principle exhibits low pressure around the flow of matter. Only here, the range of action of such a differential is long range, as the two imaginary entities together produce a *real* result, thus the negative gravitomagnetic action does not become evanescent. As the negative gravitomagnetic potential is applying radial-like, it affects a centripetal-like force which points toward the center of the circular-like motion. The deflective action has the form of a "displacement momentum" effect, which tends to deflect the rotating galaxy's celestial contents centripetal-like, without changing their kinetic energy.

Since we have an effect of negative space curvature, and that gets to project on the real world, according to what was previously noted, this should deploy on the real world through a negative reciprocal (due to $i^2=-1$ → $i=-1/i$). It should therefore affect a force that is symmetric to (5-37) of the airplane lift, copied here for easiness in reference

$$F = m\ T_{\mathrm{d}}^{+} = m\ ((1/\varepsilon_{\mathrm{g}})\ T) \qquad\qquad (5\text{-}37)\text{ or }(5\text{-}60)$$

$$F = -m\ \bar{T}_{\mathrm{d}}^{+} = -m\ ((1/\varepsilon_{\mathrm{g}})(\bar{T}\ \cos\varphi)) \qquad\qquad (5\text{-}61)$$

In the present case the interaction involves negative matter and negative gravitomagnetic field, so the actual interaction seems to refer to a form like (5-61). Here the cosine term may be in need to concern a coordinate of energy that comes due to projection from negative energy toward positive energy as per the Weinberg angle (to be considered in the next section). The reciprocal term $(1/\varepsilon_{\mathrm{g}})$ refers to $dt/\Delta x$, so in terms of units it imitates the reciprocal velocity $(1/v)$. In this case, the strength of the corresponding force (actually "potential") seems to vary with respect to the reciprocal radius $1/r$. It should therefore be less strong the closest to the center of the galaxy. Since this potential applies on top of the gravitational which varies with respect to r^2, the combination of these two independent forces (of relatively comparable strengths) would be a total force that varies roughly with respect to the radius from the center of a galaxy $(r^2) \times (1/r) \approx r$. (The approximation is rough since the two forces involved are not equal). In such way, this may account for the galaxy rotation which remains more-or-less flat as the distance from the center increases. And the negative energy (negative matter) involved seems to account for the missing (dark) matter, without a need for the existence of any additional dark-matter particles.

With some degree of speculation, the imaginary nature of the negative gravito-magnetic potential may resemble the units of the magnetic field (sec^2/m^3), which may be re-written in the form (sec^2/m)($1/m^2$) or alternatively expressed in the form (dt^2/dx)($1/dx^2$). The later expression includes two terms; an acceleration-like term (dt^2/dx), and a flux-measuring term ($1/dx^2$). The former resembles the negative reciprocal of the units of the **gravitational** field, since $d(1/x)/d(1/t)^2=(dt^2/dx)$, while the latter describes a **flux** (actually a "flow", since it takes place with respect to positive space). This could rather account for the centripetal-like transfer (flow) of negative momentum, through an infinitesimal area ($1/dA$) where A corresponds to dx^2.

On a side-note, above we have described the action of a *negative displacement gravitomagnetic force* (a force of $\bar{T}_d$) that causes the rotational curves of spiral galaxies to be flat, and applies at very long range (unlike the conventional gravitomagnetic force which was said to apply at close range). Note that it may also take intrinsic-field characteristics. It seems likely that an antisymmetric process may involve (take place through the action of) a *displacement antimagnetic force* (a force of $\bar{B}_d$). That may apply at very short range, and constitute a key reason that the strength of the gluon force between quarks is independent from distance, accounting (in part) for the flux-tubes behavior of the strong interaction (that interaction shall be considered in section 10).

Capacitive-like and Inductive-like effects:

A phase shift between the waviness in the metrics of space and time associates to effects of mechano-capacitance and/or mechano-inductance. These concern a shift forward or backward in *space* (in symmetry to a shift in *time* of the electromagnetic case). This shift provides a window for the T and $\bar{T}$ fields to interact with, and leak energy toward the opposite energy domain in each case (imaginary or real) during transitional cases, where energy is provided through a Sine or a Cosine coordinate correspondingly.

-   A mechano-capacitive-like phase shift in space (a lead) may concern effects like the displacement of the wave function in the effect of barrier penetration, or the powering of "exergonic" biochemical reactions which are involved in life-powering interactions. These seem to involve a gravitomagneto-motive potential.
-   A mechano-inductive-like phase shift in space (a lag) concerns the inertial behavior of matter, which appears to involve the gravitomotive potential, while a similar process will be discussed to engage in vortex flow creation.

These effects are considered right next, in section 6.

6
Notes on Reaction Potentials

When there is a change in the motional status of charge or matter, or when there is a change in the flux of a field (e.g. electric, magnetic, gravitational, or gravitomagnetic), any of that comes with a reaction. The *reaction* typically develops orthogonally to the *action*, meaning that either of those should have an imaginary character with respect to the other, just like a real axis and an imaginary axis are orthogonal to each other. This feat may be seen in practice, when -for instance- an equation involves the electric permittivity in the one side of it, and the magnetic permeability in the other side. In such case, either of those (but not both) should be in *reciprocal* form, signifying the imaginary origin of either one of these two terms with respect to the other. Additionally, **a change arising in the *space* domain (positive or negative), should typically come along with a reaction in the *time* domain** (positive or negative), and vice versa. For example, a change arising due to an electric field action (relating to a differential in imaginary space) comes along with a magnetic-related reaction (in imaginary time), or vice versa. This allows for a balance of energy to be maintained between the space and time domains.

Reaction potentials therefore get an action that develops in the real domain to be reacted by leaking energy toward the imaginary domain, while the process may take place in the opposite way as well. The fields that yield reaction in the opposite domain in each case, are in position to do so due to the existence of a phase shift between the curvatures of space and time. This phase shift is what provides to fields a window to apply as a field coordinate to the opposite curvature. For example, it allows for the projection of a real entity toward the imaginary world through a sine θ coordinate, or the projection of an imaginary entity toward the real world through a cosine θ coordinate, where θ is the angle of the phase shift. In this sense the reaction fields may apply in partial strength (due to the sine or cosine coordinate involved), and therefore their *reaction* is typically less effective that the corresponding *action force* which they are reacting against. This is why -for instance- the inertial impedance is less effective that the action force, so it allows for acceleration to take place. Such reaction potentials are named with the ending "motive", like for instance the **electromotive** potential. We shall here explore more reaction potentials like the **magnetomotive**, as well as the newly introduced mechanical counterparts, the **gravitomotive** and the **gravitomagnetomotive**. These will actually be figured to apply quite extensively all around us, being responsible for effects like the inertial behavior of matter, effects of vortex flow, barrier penetration, and other.

6.1 The Electromotive Potential

The electromotive force (EMF) may be denoted by the expression $\oint E \cdot d\ell$ or by the letter ε, and we shall prefer the first in order to avoid confusion with the symbol of the electric permittivity. As known, it may exercise reaction in two basic occasions, either in reacting a change in the **flow** of charge (e.g. reacting a change in current through a coil), or in reacting a change in a magnetic **flux** (e.g. through a metal ring).

(a) Reaction to a change in current.

This concerns a reaction to a change in current in an electric circuit involving a coil, like in **self-induction**. For simplicity in addressing certain concepts, we'll accept that, since an electric circuit constitutes a *loop*, we'll assume that the circuit itself corresponds to a form of basic coil (of a single turn of wiring, so it exhibits far lower inductance that an actual coil). As considered in relation to figure 5-10, the magnetic flux $\Phi_B = LI$ through that circuit loop tends to remain constant. If a change in current I is imposed to the circuit loop, this brings a change in magnetic flux through the circuit's cross-sectional area. In an effort to keep the magnetic flux Φ_B unchanged, an effect of induction takes place, which corresponds to a shift backward in time that associates to a characteristic phase lag of the current oscillation with respect to the voltage oscillation, over the circuit loop.

This phase shift gets the power consumed in trying to change the current to be split in two coordinates, a real and an imaginary one. When an external potential is applied, the real coordinate of power gets work done in changing the velocity of charges in the circuit, while the imaginary coordinate associates to energy consumed without doing actual work. While in this case the electromotive potential does not deliver work of its own, it is responsible for the impedance in trying to change the current, and that process refers to self-induction. (Note that this will later be explained to be symmetric to the way inertia applies in mechanics). In this way, the extent of the phase lag (measured in degrees of shift, from min 0 to max 90) translates to how "difficult" (inefficient) it is to carry out the change in the motional status of charge. The extent of the phase lag relates to the value of a circuit's (coil's) inductance L, where inductance refers to a shift backward in time $L=-\Delta t$. The relation describing this displacement has as follows

$$\oint E \cdot d\ell = -L\ dI/dt \tag{6-1}$$

We shall here explore certain relations that may come up by re-arrangement of terms, like for instance by transferring L to the left side of the equation, in which case the relation takes the form

$$-\oint E \cdot d\ell/L = dI/dt$$

Notice that, on the basis of units, the term $(d\ell/L)$ resembles the magnetic permeability $(1/\mu)$, so the relation (6-1) may be re-written as:

$$-(1/\mu)\oint E = dI/dt \tag{6-2}$$

What that may tell, is that a change in current I is reacted by an **inductive displacement** version of the electric field, which we shall denote by the symbol $\bar{D}-$, and refers to an expression as follows

$$\bar{D}- = (1/\mu)\,(i\,E\,\sin 0) = (1/\mu)\,(\bar{E}\,\sin\theta) \tag{6-3}$$

Here, the letter D indicates a "displacement electric field", the upper bar indicates the imaginary version of the field (therefore concerning a displacement "anti-field"), while the minus sign at its right indicates its "inductive" character, which differentiates it from the "capacitive" displacement electric field ($D=\varepsilon E$) which is responsible for the displacement of charge between a capacitor's plates that we shall describe later.

While this field is a reaction field, it should involve negative energy, and should therefore be accounted for as a projection toward the imaginary axis, where it leaks energy toward. In terms of energy this projection could be accompanied by a **sine** coordinate of an angle θ, where θ identifies to the inductive phase shift between the associated Voltage-Current oscillations. The sine coordinate concerns the energy portion retrieved (and consumed) from the circuit's energy source (the electric battery) without doing work, so it applies as a drag-like agent, which impedes the change in the motional status of charge, hence it impedes a change in current.

We shall now make a particular modification to equation (6-1), which shall be of help later in revealing a symmetry between the electromotive potential and other potentials that we shall consider in the present section. Since $d\ell$ concerns the circuit loop, it measures in terms of length dx, hence the left side of (6-1) could be expressed as $\oint E \cdot dx$. In such case we may re-write the multiplication by dx as a division by $(1/dx)$ which is mathematically equivalent. Also, we shall denote the imaginary version of the electric field by the term $\bar{E}$ (instead of $\oint E$). The new representation is shown in (6-4). The reaction field $\bar{E}$ that powers the electromotive potential changes with respect to imaginary space $idx=(-1/dx)$. That seems to describe a negative-energy (therefore imaginary) version of the electric field. As per the reciprocity of space $(1/dx)$ in the denominator of the left side, which reflects the field's imaginary character, the field "lines" of this field version should project to the real world through an inverse geometry, meaning that they should follow a local loop (instead of being radial-like and therefore looping through the far distant universe with a portion of them being imaginary, as it holds for the conventional electric field). That the field lines of this version of the electric field form loops, actually complies to existing theory. And since it concerns an imaginary version of the electric field, it seems to associate to the aforementioned **anti-electric** field.

$$\bar{E}/(1/dx) = L\ dI/dt \qquad\qquad (6\text{-}4)$$

A point of attention is that, since the field lines of $\bar{E}$ ride a loop (the circuit's loop), **its effect should apply *concurrently* throughout the whole loop**. This should be so, for same reason as, when a bicycle wheel rotates, all elements of the wheel rotate together concurrently, so any change in rotational velocity is instantaneously applicable over all elements of the wheel. What this tells, is that any reaction affected by $\bar{E}$ should be mediated **instantaneously**. Actually, such kind of fastness should be understood to comply with the imaginary character of this field version.

Furthermore, since for current it holds $I=Q/t$, the expression (6-4) may be transformed as follows

$$\bar{E}/(1/dx) = L\ dQ/dt^2$$
$$\bar{E}\ (1/L) = (1/dx)\ dQ/dt^2$$
$$\bar{E}/L = dQ(1/dt^2)\ /dx$$
$$\bar{E}/(-\Delta t) = d\Phi_Q\ /dx \qquad\qquad (6\text{-}5)$$

Here we have assumed a new notion, of the **charge flow** Φ_Q, which will actually make more sense later, when we shall consider its symmetry with other flows and fluxes. Similarly to as the electric flux corresponds to $\Phi_E = E\, dx^2 = E/(1/dx^2)$, a flow of charge may be represented as shown in (6-6). Since the denominator involves time (instead of reciprocal time) we may actually call that a regular "flow" (instead of "flux").

$$\Phi_Q = Q\,(1/dt^2) = Q/dt^2 \tag{6-6}$$

What (6-5) tells in this case, is that a change in the flow of current with respect to space is reacted by an anti-electric field per phase lag in time (referring to the electromotive potential). On a note of attention, the term "dt^2" within the expression of the flow Φ_Q may be considered to assume the existence of two dimensions of time (in a similar sense to as "dx^2" in $\Phi_E = E\, dx^2$ reflects flux through **an area** so it uses two dimensions of space). The "use" of two dimensions of time could be considered to come in contradiction to current theory which accepts a single dimension for time (the so-called "fourth" dimension on top of the three dimensions of space). However, this particular point has been addressed in section 5.4-point(3) where it was explained why the use of 2 (or even 3) dimensions of time may be acceptable. So, the denominator in $\Phi_Q = Q/dt^2$ can involve the two (out of three) dimensions of time, in particular the ones which are not phase-shifted during this interaction.

Furthermore, (6-5) may also be re-arranged so as to appear in the form (6-7) which is looser but handier in terms of comparison with other fields and potentials that we shall consider ahead.

$$
\begin{aligned}
(dx/\Delta t)\,\bar{E} &= \Phi_Q \\
(dx/L)\bar{E} &= dI/dt \\
(1/\mu)\bar{E} &= dI/dt \\
\bar{D}- &= dI/dt
\end{aligned}
\tag{6-7}
$$

Notice that since the electromotive potential associates to a shift in negative time $-\Delta t$, it resembles a magnetic type of potential. This is actually what lets it go together with a *permeability* term $(1/\mu)$, instead of a *permittivity* term that one would expect to accompany an electric field. The role of μ of course, is here deeper than simply being a measure of magnetization of materials, since its units come in Henries per meter, where the Henry is the unit of inductance. The involvement of μ accounts for the inductive shift (a shift in *negative time* $-\Delta t$) between the Voltage and Current oscillations.

Notice the difference, that in the previous section we considered the involvement of the magnetic permeability in the displacement magnetic field $H=(1/\mu)B$, where that concerned the magnetic action in the SIDEWAYS direction. While the electromotive force applies its effect along the direction of motion of charge (therefore orthogonally to the magnetic case). In this way, the electromotive force has an influence on the kinetic energy of charge (as it impedes a change of it), unlike the magnetic force which doesn't influence the kinetic energy of a charge as it only deflects its direction of motion.

On a side note, the electromotive potential brings an impeding behavior which produces a similar type of physical result (an impeding force) with its mechanical counterpart, of inertia, though they constitute different effects as the one relies on a time shift and the other on a space shift. This similarity in terms of impeding behavior will later (in section 9) be considered to resolve an illusion when it comes to a quantized behavior, where inductance will be considered to resemble the transfer of negative momentum. In particular, this will be shown to provide the illusion that the Z^0 boson of the weak interaction conveys mass, while this is not really the case.

(b) Reaction to a change in magnetic flux.

This concerns an inverse type of interaction. Instead of a change in current to be reacted by an effect of induction (in an effort to maintain the magnetic flux through the circuit loop steady), now we have a change in magnetic flux through a circuit loop to be reacted by an inductive change in current (in an effort, again, to maintain the magnetic flux through the loop unchanged).

Consider that if charge moves through a straight wire, its magnetic field is evenly formed around the wire, while if charge moves through a loop-shaped wire (like a metal ring or a coil), then all magnetic field lines are squeezed to pass through the hole of that metal ring. In such case, if an *external* magnetic field approaches that ring, the coordinate of the *external* magnetic field that crosses through the ring's hole superimposes upon any existing magnetic field through the hole due to the ring's current, and nature senses this change in magnetic flux as an "effective" change in current in the metal ring. In respond to that, a reaction (electromotive) potential is generated to react the presumed change in current. And, in the absence of an actual change in current in the first place, the electromotive potential gets to displace charge in the opposite direction, creating a "inductive reaction current".

It appears that the inductive displacement of charge in the metal ring does not exactly constitute a conventional current. It rather concerns a displacement-type of current I_d which is temporal (meaning that, even under conditions of perfect conductivity, that current would last only for as long as the change in magnetic flux takes place through the ring). This temporal displacement seems to concern an effect of the wave nature of charges (instead of the particle nature which ascribes to the conventional conduction current I). In fact, it appears that there are two kinds of displacement currents; an **inductive** displacement current like the one we are considering here and involves the electromotive potential, and a **capacitive** displacement current like the one that develops between the plates of a capacitor, which we shall consider later and will be explained to involve a different potential that we shall name magnetomotive.

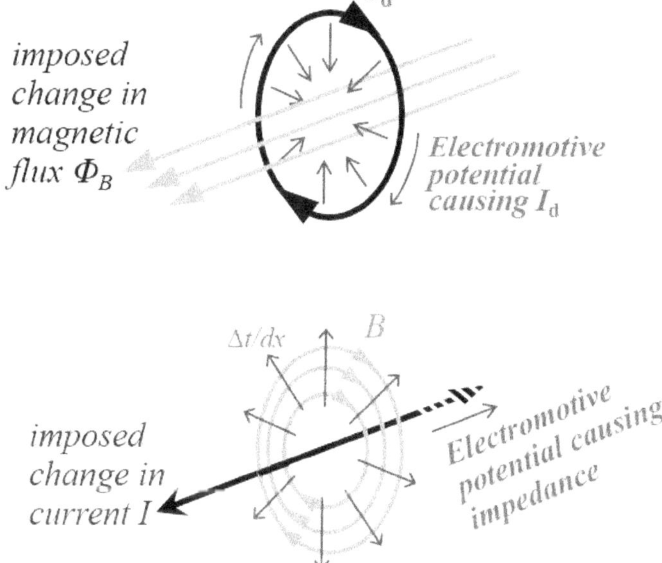

Figure 6-1: The electromotive potential reacting
a change in magnetic flux (upper) or a change in current (lower)

Figure 6-1 illustrates two symmetric effects. The upper part depicts that when there is a change in magnetic flux (three grey arrows) through the hole of a metal ring (black ring), the electromotive potential generates a current in the ring as per Lenz's law $\oint E \cdot d\ell = d\Phi_B/dt$. As described earlier, the electromotive potential is a negative energy potential, that concerns an imaginary field version of the electric field, which has loop-like field lines (instead of radial-like, as of the conventional electric field's). Even though the overall effect described by Lenz's law looks like a single interaction, the process seems to involve two different steps, as follows. (This supposition will become more clear later, when we shall describe the same concept on the effect of vortex flow in mechanics):

i.   The change in magnetic flux through the ring resembles an increase in the density of the magnetic field "lines" that passes through the hole of the metal ring. This increase in density applies as a sideways shift in magnetic field lines (bringing them closer to each other). This shifting (change of density) has the form of a negative-energy radial-like potential (see radial arrows at the upper-part of figure 6-1) which resembles the way that Bernoulli pressure develops around fluid flow in mechanics. While the magnetic flux through the metal ring changes, that corresponds to a form of **radial "displacement magnetic" field** of negative energy.

ii.   In respond to the generation of this negative energy radial-like potential, nature tries to create an opposing displacement potential. Since the latter is meant to react the radial one, it should develop orthogonally to it, therefore it should apply loop-like. That in turn, corresponds to the imaginary version of the electric field which has loop-like character. And since that concerns the **imaginary version of the electric field**, its displacement associates to displacement of anticharge, as per the term $\oint E d\ell$. The interaction involves two imaginary entities, namely the imaginary part of the wave nature of charges and a negative energy version of the electric field, it illusively appears to be much like a conventional conduction current (as if it were concerning the particle-nature of charges).

That the interaction takes place in negative energy can be also traced through the fact that the complete effect develops in an effort to react a change in magnetic **flux**, where the flux concerns an imaginary measure. That may be figured through the reciprocation involved in the units of flux. That is, when the magnetic flux is crossing an area, that is not measured in conventional terms of flow "per area" (since the flux is NOT represented as $\Phi_B = B/dx^2$), it is instead measured as a flow "times area", or equivalently, as flow per "negative reciprocal area":

$$\Phi_B = B \, dx^2 = B/(1/dx)^2 \tag{6-8}$$

Looking into the same effect from the electric field's point of view, it is like having the imaginary version of the electric field to apply on the imaginary part of charges, in an "inverse-electromotive" kind of process. The field involved in this case seems to corresponds to the projection of the imaginary version of the electric field toward the real world. That could be represented by a relation looking conceptually as follows

$$D- = (1/\mu)\,(E\cos\theta) \tag{6-9}$$

Here, the letter D denotes the displacement-electric field. As that concerns an imaginary variant of the electric field, which interacts with the imaginary portion of charges (thus both entities being imaginary), the interaction between the two resembles the interaction between the conventional electric field with conventional charge. That is why we have not placed (in terms of notation) the bar above the D and E. The minus sign indicates the inductive nature (not capacitive nature) of this displacement (hence the loop-like shape of the interaction). Furthermore, the cosine term is there to reflect the coordinate of energy due to the phase shift between voltage-current. This coordinate has a "cosine" term as it concerns leakage of energy from the imaginary toward the real domain (the opposite of what holds in self-induction).

Notice that if we were to combine in a single formula both the above two embodiments of the inductive displacement electric field (both self-induction and Lenz's law), this should look like (6-10). This is a new equation, which looks symmetric to the equation concerning the (conventional) "capacitive" version of the displacement electric field which we shall consider ahead.

$$D- = (1/\mu)\,(E\cos\theta - i\,E\sin\theta) \tag{6-10}$$

Following what has been considered above in (a) and (b), it is of interest to make a further comparison between the reaction to a change in magnetic flux (re. figure 6-1 upper part) and the previous case of self-induction which reacts a change in current (lower part of the same figure):

-   In the lower part of the figure, consider that the straight wire segment is (in any case) part of a bigger closed-circuit loop. Since in the case of a loop all magnetic field lines cross through the loop, the setting is pretty-much similar to that of the upper part of the figure, in terms of tending to keep the magnetic flux through the loop steady.
-   In the other hand, since self-induction reacts a change in the "flow" of charge, while Lenz's law reacts a change in the "flow" of magnetic flux, these two processes may be considered "orthogonal" to each other. This orthogonality lies in the fact that self-induction seems to react a change in the motional status of the real portion of charges, while Lenz's law seems to apply its reaction through the motional status of the *imaginary* portion of particles' wave nature, which has freedom to displace in an orthogonal

direction. So, in Lenz's law we have an inverse process taking place, since the change in current here is not the "cause" but is rather the "effect" of the process. Furthermore, in self-induction we have energy leaking from the real toward the imaginary domain, while in Lenz's law we have energy leaking from the imaginary domain toward the real domain by bringing physical outcome (displacement of current) thanks to the relation $i^2=-1$. The negative energy of the later also necessitates the use of the permittivity and permeability terms, in relation to what was described in section 5.3.

Furthermore, just like it was done for Lenz' law, self-induction could also be approximated through a two-step process, where:

(i) The change in current causes a change in magnetic field (around the wire) as per Ampere's law. Since the magnetic field corresponds to equipotential lines/surfaces, a change in magnetic field corresponds to a radial-like capacitive differential $\Delta t/dx$ (see radial-like arrows in figure 6-1 lower part) which applies in a similar way to as a fluid flow generates a Bernoulli pressure around it in mechanics. If that capacitive differential is evenly distributed (co-axially) around every segment of the wire, it does not produce a net magnetic force. However, since the wire eventually forms a loop (a circuit loop, or the loop of the turns of a coil), the magnetic field is not coaxial, as all the magnetic field lines are squeezed to pass through the loop. Since it is not coaxial it corresponds to a potential, though that potential cancels out over the whole circuit loop so it does not deliver work.

(ii) In respond to the corresponding deformation of the magnetic field over the loop (not shown in the figure), nature tries to create an opposing process, where that process would deliver a reaction to the source that provides the power for that change. As mentioned, the term $((-1/\varepsilon_0)\bar{B})$ resembles the imaginary variant of the electric field (as per $\oint E$), which has loop-like character. Since the opposing action points exactly opposite to the direction of acceleration of charges, it only reacts the change without causing displacement motion of its own. This is what brings the effect of self-induction (in an inertial-like way).

6.2 The Gravitomotive Potential (in inertia & vortex flow)

Same as with its electromotive counterpart, a gravitomotive effect also exhibits itself in two basic ways; either in reacting a change in the motional status of matter (which refers to what we perceive as "inertia"), or in reacting an effective change in gravitomagnetic flux (where that will be discussed to concern vortex flow). In the first case the gravitomotive potential only reacts, while in the second case it is able to generate displacement.

(a) Reaction to a change in the motional status of matter (inertia)

As of today, there is no intuitive explanation of how inertia applies, there is no equation to describe it, neither there is a particular field associated to it. One may loosely assert that "action equals reaction", however this assertion is not accurate, because if it were, the object would not be able to accelerate at all. The inertial reaction should therefore be somewhat weaker to the action force, while it could also be of different essence than the force that drives the acceleration. It turns out that the latter seems to be the case, with the inertial reaction attributed to a **gravitomotive potential** (GMP).

While in the back-electromotive case the phase shift was associated to a lag in time (Δt), in the case of inertia a corresponding phase shift appears to associate to a lag in space (Δx). That lag in space actually corresponds to the particular phase difference between the Space and Time curves shown in figure 1-1. That phase shift may be represented in degrees of angle, here discussed to match the "**weak mixing angle**" of the electroweak interaction. This shift gets the power consumed in trying to change the velocity of a material object to be split in two coordinates, a real and an imaginary. The real coordinate of power does work in changing the velocity of the object, while the imaginary coordinate of power gets energy consumed (in reacting the change) without doing actual work. **The part of the power associated to the imaginary coordinate applies as an "inefficiency" in changing the velocity of a material object, where that corresponds to what we perceive as inertia**. In this sense, the extent of the phase lag (measured in degrees of shift, from min 0 to max 90) translates to how "difficult" (inefficient) it is to carry out the change in the motional status of matter. In relation to material to be presented later through figure 7-1, for the case of conventional inertia this shift corresponds to the weak mixing (the Weinberg) angle, estimated at 30 degrees. Furthermore, the existence of this phase shift throughout the universe (as depicted in figure 1) corresponds to the so-called **Higgs** field that is said to permeate the vacuum. In this sense, the phase shift can be understood to apply as an inductive-like field that we may call "mechano-inductive" in order to distinguish it from the inductive one of electronics. As mentioned, the difference between these two types of phase shifts lies in the fact that the mechano-inductive concerns a shift in space, while the inductive (of electronics) concerns a shift in time. And just like the GMP is responsible for the inertial resistance in changing the motional status of matter, it is also responsible the property of **momentum** that materials exhibit, since momentum corresponds to the resistance in changing (either reducing or increasing) the motional status of moving matter. A quantitative treatment of the GMP is symmetric to the treatment of the back-electromotive potential. The corresponding field could be denoted as g_{im} or $\bar{g}$, or loosely as $\oint g$ and we may approximate the GMP via a self-induction-like law as follows:

$$\oint g \, d\ell_t = -L_g \, dp/dx \qquad (6\text{-}11)$$

The constituents of this equation and their units have as follows:

- The term $\oint g$ corresponds to an imaginary (negative energy) version of the gravitational field, which has loop-like field lines (same as in the electromotive case). Its core units are same with the ones of the conventional gravitational field, though they should go together with an imaginary term i, where that may apply as an equivalent of a reciprocation in units.
- The quantity $d\ell_t$ corresponds to a time loop, over which the phase shift develops, as we shall consider below. Its units can be "seconds", and since it concerns time, the notion of spatial extend (over which it deploys) is irrelevant. This means that there is no issue even if the loop is immense in spatial terms.
- The quantity L_g is symmetric to induction L in electronics, in the present case it concerns the shift in space Δx between the wavinesses of the space and time metrics illustrated in figure 1-1 (in terms of angle, this identifies to the Weinberg angle).
- The p corresponds to a particle's momentum. For a particle's wave nature, it holds $p=(h/\lambda) = (h/dx)$, which is symmetric to the expression of the current $I=Q/dt$ in electronics. Notice that for a typical elementary particle like the electron both h as well as Q have quantized values. In that sense, **the quantity $dp/dx = d(h/\lambda)/dx$ can be loosely measured in units corresponding to dh/dx^2 in certain analogy to as dI/dt is measured as dQ/dt^2.**

Transferring L_g to the left side of the equation, this becomes

$$-\oint g \, d\ell_t / L_g = dp/dx \qquad (6\text{-}12)$$

Notice that, based on units, the term $(d\ell_t / L_g)$ resembles the permeability-like term $(1/\mu_g)$, where $\mu_g = -L_g/dt$, and $L_g = -\Delta x$ is the mechano-inductance which was introduced above. In this sense, the relation (6-11) may be re-written as:

$$-(1/\mu_g)\oint g = dp/dx \qquad (6\text{-}13)$$

What this may tell, is that a change in momentum p is reacted by a mechano-inductive **displacement** version of the gravitational field. We shall denote the corresponding field version by the letter $\bar{g}_{d-}$, where the subscript "$_d$" indicates its displacement field character, the upper bar indicates the imaginary version of the field (therefore reflecting a negative-energy-field), and the minus sign at its right side indicates its "inductive-like" displacement nature (which differentiates it from the "capacitive-like" displacement gravitational field, which we shall later associate to the effect of barrier penetration). This should have as follows

$$\bar{g}_{d-} = -(1/\mu_g)\,(i\,g\,\sin\theta) = -(1/\mu_g)\,(\bar{g}\,\sin\theta) \qquad\qquad (6\text{-}14)$$

As this field is a reaction field, it should involve negative energy, and should therefore be accounted for as a projection toward the imaginary axis, where its energy associates to. In terms of energy this projection could be accompanied by a **sine** coordinate of an angle θ, where this angle identifies to the phase shift between the Space and Time metric wavinesses (shown in figure 1.1). In this sense **the sine coordinate concerns the energy portion consumed from the acceleration source without doing work, thus it is responsible for a drag-like behavior which impedes the change in the motional status of matter**.

Gravitomotive case	Electromotive case
$\oint g\cdot d\ell_t = \theta\Phi_T/\theta x$	$\oint E\cdot d\ell = \theta\Phi_B/\theta t$
and since $\Phi_T = L_g\,p$, we get	and since $\Phi_B = LI$, we get
$\oint g\,d\ell_t = -L_g\,dp/dx$	$\oint E\cdot d\ell = -L\,dI/dt$
or $-(1/\mu_g)\oint g\,d\ell_t = dp/dx$	or $-(1/\mu)\oint E\cdot d\ell = dI/dt$
and since $p=h/\lambda$, we get	and since $I=Q/t$, we get
$\oint g\,d\ell_t = -L_g\,dh/dx^2$	$\oint E\cdot d\ell = -L\,dQ/dt^2$
$\bar{g}/(1/dt) = L_g\,dh/dx^2$	$\bar{E}/(1/dx) = L\,dQ/dt^2$
$\bar{g}\,(1/L_g) = (1/dt)\,h/dx^2$	$\bar{E}\,(1/L) = (1/dx)\,dQ/dt^2$
$\bar{g}/L_g = [dh(1/dx^2)]/dt$	$\bar{E}/L = [dQ(1/dt^2)]/dx$
$\bar{g}/(-\Delta x) = d\Phi_h/dt$	$\bar{E}/(-\Delta t) = d\Phi_Q/dx$
or	or
$(dt/\Delta x)\,\bar{g} = d\Phi_h$	$(dx/\Delta t)\,\bar{E} = \Phi_Q$
$(d\ell_t/L_g)\,\bar{g} = dh/dx^2$	$(d\ell/L)\bar{E} = dQ/dt^2$
$(1/\mu_g)\,\bar{g} = dp/dx$	$(1/\mu)\bar{E} = dI/dt$
$\bar{g}_{d-} = dp/dx$	$\bar{D}- = dI/dt$

Table 6.1:   Gravitomotive vs. Electromotive relations

We shall now consider a particular modification to equation (6-11), similar to the one we did previously for (6-1). For convenience we place these two modifications side-by-side in Table 6.1, to display the symmetry. The left side of the table concerns the Gravitomotive case while the right side of the table concerns its Electromotive counterpart that we presented earlier.

Since $d\ell_t$ concerns a time loop, it represents a measurement of dt, so the left side of (6-11) may be loosely re-written as $\oint g \cdot dt$. In such case we may switch the multiplication by dt to a division by $(1/dt)$ which is mathematically equivalent. This representation comes along with the fact that the reaction field that powers the gravitomotive potential corresponds to a negative-energy (therefore imaginary) version of the gravitational field. Due to its imaginary character which comes along with a phase shift, its projection to the real world should come through an inverse geometry, with its field lines forming loops (instead of being radial-like as of the conventional gravitational field). Being an imaginary version of the gravitational field, this seems to correspond to a **negative-gravitational** field, which we shall denote by the symbol $\bar{g}$. The relation (6-12) may therefore be expressed also as follows

$$\bar{g}/(1/dt) = L_g \, dh/dx^2 \tag{6-15}$$

Unlike the electromotive case where the field lines of the imaginary version of the electric field were following a loop in space (the circuit loop), in the mechanical case (concerning inertia) the loop is in time. In relation to material to be considered in section 7 such loop seems to ride the axis of the time metric that its waviness concerns. So, it may be imagined to refer to the loop-like axis XX' of the waviness of the metric of time, depicted in figure 1-1. So that should be understood to concern a loop in time through the very distant universe (where, as mentioned earlier, the spatial extend is irrelevant). The corresponding phase shift between the waviness in the metric of time with respect to the waviness in space metric, is therefore responsible for bringing the inertial impedance. Furthermore, since for momentum it holds $p=h/dx$, the expression (6-15) may be re-written as follows

$$\bar{g}/(1/dt) = L_g \, dh/dx^2$$
$$\bar{g} \, (1/L_g) = (1/dt) \, dh/dx^2$$
$$\bar{g}/L_g = [dh(1/dx^2)]/dt$$
$$\bar{g}/(-\Delta x) = d\Phi_h/dt \tag{6-16}$$

Here, we have assumed a new notion, of **matter flow** Φ_h. Similarly to as the electric flux corresponds to $\Phi_E = E \, dx^2 = E/(1/dx^2)$, a flow of matter may be represented as a flow per space by a relation of the form

$$\Phi_h = h\,(1/dx^2) = h\,/dx^2 \qquad\qquad\qquad (6\text{-}17)$$

Since here the division by dx^2 concerns an area of ordinary "space" (instead of "negative reciprocal space") we may refer to it as a "flow" (instead of "flux"). So, what (6-16) tells, is that a change in the flow of matter per time ($d\Phi_h/dt$) is reacted by a (loop-like) negative gravitational field $\bar{g}$ with respect to the phase shift in space $-\Delta x$ (the Higgs field's phase shift).

Furthermore, the relation (6-16) could also be re-arranged to appear in the form (6-18), involving the inductive displacement negative gravitational field $\bar{g}_{d-}$. This form is rather looser but handier in comparing with other fields and potentials being considered here (as for instance (6-7)).

$$(dt/\Delta x)\,\bar{g} = -d\Phi_h$$
$$(d\ell_t/L_g)\,\bar{g} = -dh/dx^2$$
$$(1/\mu_g)\,\bar{g} = -dp/dx$$
$$\bar{g}_{d-} = -dp/dx \qquad\qquad\qquad (6\text{-}18)$$

According to that, the gravito-motive potential applies as a "negative inductive displacement gravitational field", over the loop of the waviness of the time metric (the axis XX' of figure 1-1). The way that $\bar{g}_{d-}$ mediates the inertial reaction potential, is similar to conveying negative momentum. A relevant point of significance, is that since such a field is imaginary, it should rather apply its reaction at superluminal fastness. This condition is in fact satisfied by default since the phase shift should take place throughout the whole axis XX', which means that all elements of the loop are noticing this shift *concurrently* (just like when a bicycle wheel that rotates, where all elements of the wheel rotate together concurrently, so any change in rotational velocity is instantaneously applicable over all elements of the wheel). This feat of instantaneity is crucial when affecting inertia, as, if there were a time-delay in exercising inertia, that would create a problem during rotational effects, where a delayed inertial reaction could be delivered toward a different direction (due to rotation during the time elapsed until the reaction applies), and that does not seem to be happening.

On a side note, the description of above concerns conditions of unitary spacetime (meaning classical conditions where the reference frame refers to specific values of the time and space metrics which prevail in our cosmic neighborhood). In section 9 we shall consider similar conditions applied on non-unitary spacetime (meaning that the metrics of time and space have oscillatory values), with implications in the electroweak interaction.

(b) Reaction to a change in the gravitomagnetic flux (vortex flow)

Reasons of symmetry suggest that the gravitomotive potential should also be involved in an interaction that resembles Lenz's law. Such effect of reaction should induce vortex flow like in the case of the whirling of water that is draining down the sink (figure 6-2, *right* part), through a process which is symmetric to the one inducing a current in a metal ring when there is a change in magnetic flux through the ring (figure 6-2, *left* part).

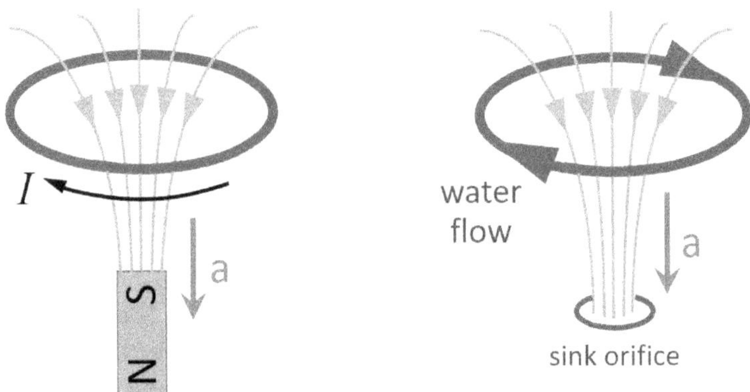

Figure 6-2: Symmetry between an induced current in a metal ring when there is a change in magnetic flux through the ring (left), and its mechanical counterpart of vortex flow (right).

To approximate the particulars of the interaction, we may start by preparing a relation that would be symmetric to Lenz's law:

$$\oint g \cdot d\ell_t = -\, \partial \Phi_T / \partial x \qquad\qquad (5\text{-}39) \text{ or } (6\text{-}19)$$

The term $\oint g$ (which could more appropriately denote as $\bar{g}$ or g_{im}) corresponds to a loop-like variant of the gravitational field which associates to an inductive-like phase shift in *space* taking place over a time loop $d\ell_t$, where that loop in time coincides to the spatially circular path of the vortex flow. (Notice the symmetry to the electric counterpart of Lenz's law, where an inductive shift in time takes place over a loop in space, the circuit's loop). As mentioned in section 5.4, the value of flux Φ_T could be figured on the basis of symmetry to its magnetic counterpart $\Phi_B = LI$. In this case, it should hold:

$$\Phi_T = L_g \, p \qquad\qquad\qquad (5\text{-}55) \ \text{ or } \ (6\text{-}20)$$

Here, the L_g concerns the shift in space as per the Weinberg angle, and p corresponds to momentum as per the wave nature. If we plug in (6-20) to (6-19), we get (6-12). But of course, (6-19) applies under a different geometric setting, inducing a loop-like potential which generates displacement motion over a loop-like path (vortex) where matter has freedom to move.

Other than the potential which generates the loop-like displacement (vortex) motion, it seems that the interaction also involves another potential as well, which develops radially (centripetal-like) with respect to the axis of symmetry of the vortex flow. In that case we need to figure what holds for that other potential, and how does it get reacted too. It turns out that, even though the overall effect looks like a single interaction, it seems to consist of two steps:

(a) The ***first step*** concerns the descent of water toward the sink's hole due to gravity, as shown in figure 6-3, left part. Notice a difference with respect to the electric case described earlier, that here the effect concerns a change in flow of *matter* (not a change in *flux*, a gravitomagnetic flux), while in induction we had a change in magnetic *flux* (not a change in flow of *charges* as a current). As the water flows downwards, the fluid flow causes a reduction in the sideways Bernoulli pressure (see centripetal-like arrows in the figure), which corresponds to a radial-like differential $\Delta x/dt$. As per its units, this tendency applies as a "displacement-velocity" $v_d = \Delta x/dt$. This is a negative energy effect, and as noted earlier, under conditions of negative energy, the conventional velocity v (or even a displacement velocity v_d) is perceived as a potential. In this case we have a "displacement velocity", associating to such a "displacement" potential, and as the water accelerates downwards, this tendency and displacement potential become stronger. (On a side-note, one may notice that on top of the above effect there is an additional centripetal "suction-like" force which applies as the water that leaves the sink needs to be replenished due to the external air pressure, but that is an independent effect so it won't be considered here).

Notice that there may be a complementary way of interpreting the process shown in figure 6-3 left part. This is possible due to the inversion of the parameters of space and time with respect to the electric interaction, where the *velocity* term is reciprocated in $g=(1/v)T$ with respect to its electric counterpart $E=(v)B$, were that allows for the gravitational field to take the active role. As portions of water are let to fall and accelerate toward the sink's hole, during their fall they should **feel more and more weightless**.

This happens due to the principle of equivalence. It's the same principle that applies in astronauts' training inside an airplane that follows a parabolic flight, where a free-falling person feels weightless. In this case, the portions of water which are falling may be considered to experience a change (reduction) in gravitational field strength, which in turn induces the radial-like gravitomagnetic field (reflecting a change in pressure), the equipotential surfaces of which has a shape like that of the funnel formed by the whirling of water (see figure 6-3 right part).

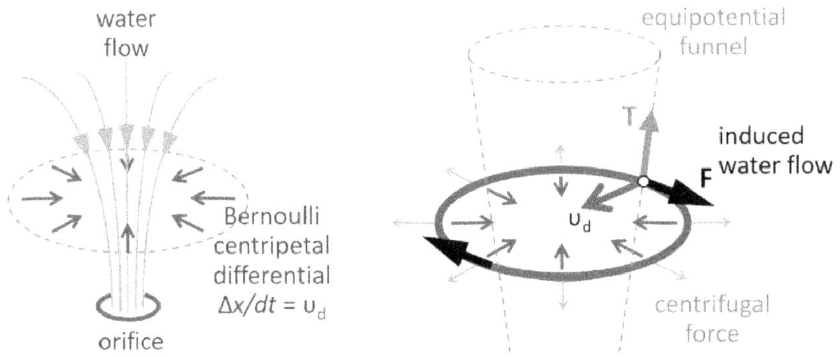

Figure 6-3: Whirl effect, in two steps.
Left: Falling water causes a gradual centripetal Bernoulli potential
Right: The centripetal potential induces circular flow. That actually causes a centripetal force which opposes the centripetal one

(b) The **second step** concerns nature's respond to this radial displacement differential. Imagine that IF there was water moving circularly already (whirl-like) that would come along with a gravitomagnetic field around the flow, looking much like the magnetic field of a current of a metal ring. Just like the current in a metal ring has its magnetic field "lines" squeeze through the ring, here the gravitomagnetic field "lines" squeeze through inside the whirl. And since they are more dense inside the whirl than outside of it, that should cause a centripetal Bernoulli potential. But here we have a sort of transitional case. As shown in figure 6-3 right part, while the water accelerates down toward the sink's orifice, the Bernoulli potential becomes stronger, so the field T changes strength and its shape takes the form of equipotential cones, much like the funnel that the water creates. And as the water keeps falling down, this resembles a change in gravitomagnetic flux $\partial\Phi_T/\partial x$. This generates a loop-like gravitomotive potential $\oint g\,d\ell_t$ as described by the relation (6-19), which induces the cycling of water (the vortex flow), very much like $\oint E d\ell$ creates a displacement current in a metal ring in the electromotive case. This is a negative energy effect. Notice that the induced cycling of water creates a centrifugal force, which reacts the centripetal potential, and this is why the effect is self-driving in order to minimize the potential energy in the complete system.

As per the above, the vortex flow, which is an effect of reaction, does not constitute a conventional flow. It rather concerns a loop-like "displacement-momentum" p_d, which seems to be drawn from particles' negative energy portion. However, since the corresponding interaction seems to involve two negative energy entities (the negative energy field version of the gravitational field and the negative portion of electrons' wave nature), the end result resembles classical flow of matter as per $i^2=-1$.

The combination of the induced circular (vortex) motion along with the falling of water through the orifice, creates the combined whirl-like flow. The effect of circular (vortex) flow is degenerate in terms of direction of rotation, as either clockwise or counterclockwise reaction works fine in creating the centrifugal forces that tend to minimize the initial change imposed. Actually, the whirling would normally take place in a specific direction of turn for fluids made of positive matter particles, and in opposite direction of turn for fluids made of matter of negative sign. But negative matter does not reach form in the classical side of the universe, as both matter and antimatter constitute "positive matter" as they are gravitationally attractive. So, this causes the degeneracy and indifference in terms of the direction of the whirl-like rotation, letting the direction of turn depend on other smaller influences, like the Coriolis force.

On a side-note, the displacement momentum may be of two kinds. Either mechano-**inductive** displacement like the one currently considered in vortex flow, or mechano-**capacitive** displacement which will be explained later to become involved in the effect of barrier penetration. In all corresponding transitional interactions, reference is made to the imaginary (negative energy) variants of fields. Notice that these imaginary versions refer to the grey arrows appearing in the "hypothetical side views" shown in figures 5.4 and 5.5. As mentioned then, for reasons of simplicity these representations don't depict what holds at the marginal portions of the field lines next to the nodes (the node at the particle, and the node at distant universe). A main reason that these portions have not been depicted in the illustrations is because the representation of a gravitational field line concerns only the curvature in the time metric, and the representation of an electric field line concerns only the curvature in the (negative reciprocal) space metric. If we were to incorporate both these curvatures (of time and space) in a single 2-D drawing, then an additional axis would be in need, as shown in figure 6-4.

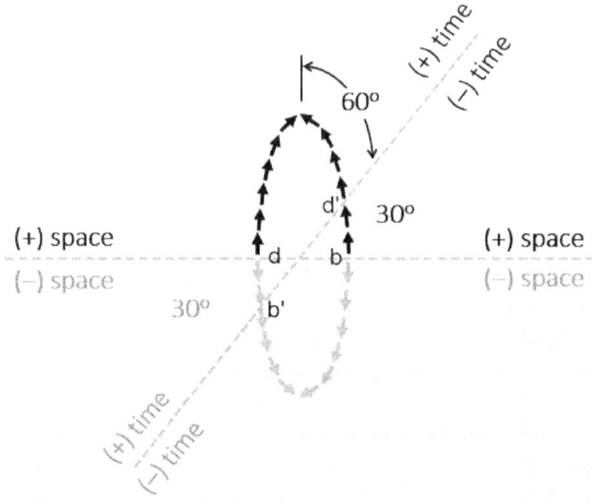

Figure 6-4: Complete hypothetical generic field line, passing from areas of positive and negative Time and Space curvatures

This figure depicts a generic field line (which could concern a gravitational or an electric field line). Notice that it passes through areas where both the space and time metrics are positive (arc d – d'), or only space is positive (arc d' – b), or both space and time metrics are negative (arc b – b'), or only time is positive (arc b' - d). Notice that these particular points (d, d', b, b') refer to same ones of figure 1-1. Notice also, that since this particular field line is generic, the "arrows" are only schematic without a tight particular meaning. In the particular way they are drawn in terms of nodes, they rather seem to match to the curvature of space instead of the curvature of time. Actually, instead of all arrows having the same direction, an alternative option has been chosen here, where the arrows have a specific direction (inclination) from the particle until half-way to distant universe, and an opposite direction (inclination) from there until the very distant universe. This is just a different representational option. The parts of the field line where either of the metrics is positive and the other is negative, typically refer to that part of the field line that engages in effects of reaction through a sine and cosine coordinate depending on the case, as considered in effects of the present section. The 60° angle shown (90° minus the Weinberg angle of 30°) seems to affect orthogonal balancing fields (like the magnetic being orthogonal to the electric).

6.3 The Magnetomotive Potential

The magnetomotive effect seems to apply in two basic ways just like its electromotive counterpart does, one way involving a change in field flux and another involving a change in the motional status of charge. A key difference with respect to the electromotive case, is that here it's the magnetic field (instead of the electric) that is accounted for in relation to *space* ($\oint B \cdot d\ell$), and it's the electric flux (instead of the magnetic) that changes per time. The swapping of the parameters of space and time constitutes a significant difference, as because of it the interaction now requires the involvement of the permittivity ε and/or permeability μ terms. It actually appears, that the units of $\varepsilon = +\Delta t/dx$ and $(1/\mu) = dx/-\Delta t$ are such, that the involvement of these terms applies as an indirect way of turning back to the alternations taking place with respect to the parameters of space and time as they hold in the electromotive case. Subject to that, the magnetomotive potential becomes involved in capacitive effects, as follows.

(a) Reaction to an alternation in electric field flux (capacitor)

The swapping in the parameters of space and time in the magnetomotive interaction (6-21) makes the field alternations subtler than in the electromotive case, since the electric permittivity term involves capacitance C (as it is measured in units of Farad/meter), and the magnetic permeability term involves inductance L (as it is measured in Henry/meter). The interaction is described by Maxwell's fourth law:

$$1/\mu_o \oint \boldsymbol{B} \cdot d\ell = \varepsilon_o \, \theta\Phi_E/\theta t \qquad\qquad\qquad \text{(5-4) or (6-21)}$$

The involvement of ε and μ allows to re-write the above in the form (6-22), since it holds $H = (1/\mu_o)B$ and $D = \varepsilon_o E$.

$$\oint \boldsymbol{H} \cdot d\ell = \theta\Phi_D/\theta t \qquad\qquad\qquad\qquad (6\text{-}22)$$

As capacitance and inductance may be complex entities (involving a real as well as an imaginary part), this also allows for the displacement electric and displacement magnetic fields to be complex entities as well, since it holds $D = \varepsilon(E\cos\theta - iE\sin\theta)$ and likewise it should hold for the displacement magnetic field as well, through a form conceptually like $H = (1/\mu)(-B\cos\theta + iB\sin\theta)$. The existence of two coordinates (cos and sin) provides to the magnetomotive field and potential the ability to engage under two ways of interaction which are orthogonal to each other. In particular, **the *imaginary* coordinate seems to**

account for the displacement of charge (in the form of displacement current), while the *real* coordinate seems to concern the involvement of the magnetic field in interactions in positive energy.

An example of the displacement of charge via the magnetomotive effect concerns the displacement of current through the plates of a capacitor. This is an effect of reaction, so it takes place only for as long as the electric flux changes. Since it concerns an effect of reaction, it should involve negative energy, thus it should concern an imaginary version of the magnetic field. The latter assumption is actually supported by the fact that (i) the permeability term is reciprocated, which is what holds for the imaginary entities that get to project to the real world, (ii) the corresponding potential damps out quickly in the classical world as it becomes evanescent, which is what generally holds for negative energy fields/waves. This is why the range of action of this effect is very short, hence it finds uses only in cases of close proximity, like between a capacitor's plates (which are very close to each other), or in wireless chargers (where the device to-be-charged needs to be close to the charging source). Notice that this effect will later be shown to be symmetric to the effect of barrier penetration in mechanics. (iii) The representation of the magnetomotive potential $(1/\mu_o \oint \boldsymbol{B} \cdot d\ell)$ involves the term "times $d\ell$" instead of a conventional "with respect to $d\ell$". The multiplication by $d\ell$ is mathematically equivalent to a division by $(1/d\ell)$, so the interaction seems to represent a change taking place "with respect to $(1/d\ell)$", where the reciprocity indicates the imaginary essence too.

This reciprocity also comes along with an inverse type of geometry, so in the magnetomotive case the imaginary variant of the magnetic potential involved should have **radial-like** field lines, instead of the conventional loop-like field lines of the magnetic field. One may recall that an analogous inversion in geometry also applies to the imaginary variant of the electric field, where that one has loop-like field lines instead of the radial-like field lines of the conventional electric field.

That the interaction involves imaginary variants of the fields, should imply that it interacts with the imaginary part of the charges' wave nature, where the interaction of two imaginary entities brings a negative real physical outcome. And since it concerns the particles' wave nature, the displacement of charge through the capacitor's plates **is symmetric to the effect of frustrated total internal reflection**, here materializing in the time domain, (through the capacitive shift in time), which is what differentiates a capacitive "displacement current" from a conventional "conduction current". Furthermore, the imaginary nature of the interaction implies that it should bear the characteristics of a **tunneling** process, in particular a tunneling in time. This complies to the notion that the displacement takes place loop-like throughout the circuit loop so it is mediated instantaneously throughout the loop, just like all elements of a bicycle's wheel move concurrently (instantaneously with respect to each other).

Let us now look to the same interaction (6-21) from the electric field's point of view. The involvement of the electric "flux" (instead of "field") indicates that what we have on hand is an inverse curvature (therefore imaginary) effect. This can be figured through the fact that the electric flux that crosses an area is not measured in terms of flow per area (the flux is not represented as $\Phi_E=E/dx^2$), it is instead measured as a flow per "negative reciprocal" area $\Phi_E=E/(1/dx^2)=Edx^2$. This is also in-line with the fact that the interaction is of short range in the real world.

Notice that since for the electric flux it holds $\Phi_E=Q/\varepsilon_o$, the right side of Maxwell's fourth law may turn to $\varepsilon_o\theta\Phi_E/\theta t \rightarrow \theta Q/\theta t$, where that corresponds to current $I=Q/t$. In that sense, Maxwell's fourth law seems to reduce to equation (6-23) which looks pretty much similar (but not identical) to Ampere's law.

$$(1/\mu_o)\oint B\cdot d\ell = I_d \qquad\qquad (6\text{-}23)$$

The reason it is not identical, is because (6-23) involves a displacement current I_d (instead of a conduction current I that Ampere's law does) where that arises from the change in electric flux $\varepsilon_o\theta\Phi_E/\theta t$. The "displacement" character comes from the presence of the electric permittivity term $\varepsilon_o=C/dx$ where that involves capacitance, and capacitance concerns a phase shift $+\Delta t$. Consequently, the relation $\varepsilon_o\theta\Phi_E/\theta t \rightarrow \theta Q/\theta t$ should describe charge being "displaced in the time domain" by $+\Delta t$ (instead of being moved conventionally with respect to dt). The degree of displacement relates to the shape, size, and proximity between a capacitor's plates, which define the value of capacitance $C=+\Delta t$.

The relation (6-23) allows to explain in intuitive terms the way that the capacitive shift in the time domain takes place. Since for current it holds $I=Q/t$, and for the charge it holds $Q=CV$, these relations should apply for the displacement current as well, in which case

$$1/\mu_o\oint B\cdot d\ell = I_d \;\rightarrow$$
$$1/\mu_o\oint B\cdot d\ell = Q_d/dt \;\rightarrow$$
$$1/\mu_o\oint B\cdot d\ell = C\,V_d/dt \qquad\qquad (6\text{-}24)$$

If we transfer the permeability term $\mu_\mathrm{o}=(-\Delta t/dx)$ to the right side, that side becomes $(-\Delta t/dx)\,C\,V_\mathrm{d}/dt$, where V_d concerns displacement-voltage. We may loosely cancel out the numerator Δt and the denominator dt since they both concern units of time (even though the first concerns a *shift* in time), and get:

$$\oint \boldsymbol{B}\cdot d\ell = C\,V_\mathrm{d}/dx \qquad\qquad (6\text{-}25)$$

Notice the symmetry between this expression of the magnetomotive potential, and its electromotive counterpart (6-1) which reads $\oint \boldsymbol{E}\cdot d\ell=-L\,dI/dt$.

On a side note, the transition from (6-24) to (6-25) may provide a hint that the right side of (6-1) might have also involved a similar cancelation looking like $(-\Delta x/dt)\,L\,dI/dx$, where that refers to $(\mu_\mathrm{g})\,L\,dI/dx$. Then, by moving the mechano-permeability to the left side, (6-1) would have taken the following form which seems more complete

$$(1/\mu_\mathrm{g})\,\oint \boldsymbol{E}\cdot d\ell=-L\,dI/dx \qquad\qquad (6\text{-}26)$$

Reverting to (6-25), we shall go a step further, and transfer the capacitance C from the right side to the left

$$(1/C)\,\oint \boldsymbol{B}\cdot d\ell = V_\mathrm{d}/dx$$
$$(d\ell/C)\,\oint \boldsymbol{B} = V_\mathrm{d}/dx$$
$$(1/\varepsilon_\mathrm{o})\,\oint \boldsymbol{B} = V_\mathrm{d}/dx \qquad\qquad (6\text{-}27)$$

Here, the existence of the electric permittivity ε_o (instead of the permeability μ_o) next to the imaginary version of the magnetic field $\oint \boldsymbol{B}$ is not by coincidence. Since the interaction applies in the imaginary domain, it concerns an opposite viewpoint than what Ampere's law does, thus the shift in time relating to the magnetic field is capacitive instead of inductive. (Notice the symmetry between the relations (6-27) and (6-2) of the electromotive case which reads $-(1/\mu_\mathrm{o})\oint \boldsymbol{E}=dI/dt$). That we now have ε_o (instead of μ_o) next to the magnetic field is also fine in the sense that either of these terms have same units, they only differ in the sign of the involved shift in time now being $+\Delta t$ (instead of $-\Delta t$). What (6-27) tells, is that a change in voltage is reacted by a capacitive **displacement** variant of the magnetic field, where the shift in time effectively concerns a tunneling effect across the capacitor's plates.

In the magnetomotive case, the capacitive-displacement imaginary version of the magnetic field may be expressed as follows

$$\overline{H}+ = (1/\varepsilon)\,(-\,i\,B\,\sin\theta) = (1/\varepsilon)\,(-\overline{B}\,\sin\theta) \qquad\qquad (6\text{-}28)$$

Here, the letter H indicates a "displacement magnetic field", the upper bar indicates the imaginary version of the field (therefore reflecting an "anti-field"), while the plus sign at its right indicates its "capacitive" character, which differentiates it from the rather more conventional "inductive" displacement magnetic field $H=(1/\mu)B$ which becomes involved in the deflection of a moving charge. Here, one may notice the symmetry between (6-28) and (6-3) of the electromotive case. While this field is a reaction field, it should involve opposite energy, and should therefore be accounted for as a projection toward the real axis where it leaks energy toward. In terms of energy this projection could be accompanied by a **negative sine** coordinate of an angle θ, where θ identifies to the capacitive phase shift between the associated Voltage-Current oscillations. The sine coordinate concerns the energy portion retrieved (and consumed) from the circuit's energy source (the electric battery) without doing work, so it applies as a drag-like agent, which impedes the change in the motional status of charge, hence it impedes a change in current.

Reverting to (6-27), if we transfer the term dx from the right side to the left, then we may get an Amperes'-law-like expression which concerns how the imaginary version of the displacement magnetic field relates to the displacement of voltage (instead of current).

$$(1/\varepsilon_0)\,\oint B = V_{\mathrm{d}}/dx$$
$$(1/\varepsilon_0)\,\oint B\cdot dx = V_{\mathrm{d}}$$
$$(1/\varepsilon_0)\,\oint B\cdot d\ell = V_{\mathrm{d}} \qquad\qquad (6\text{-}29)$$

The latest relation could actually be written in reverse since the displacement voltage V_{d} is not the "cause", but is the "effect" of the process. Furthermore, the displacement *lag* in voltage (with respect to current) may be alternatively treated as a *lead* in current (with respect to voltage) which is what capacitance is conventionally referred to associate with.

To look into the role of the change in electric flux $\varepsilon\partial\Phi_E/\partial t$ in the interaction, it is like approximating the interaction from the electric field's point of view. Similar to as in (6-3) we considered the involvement of an "inductive-displacement" version of the electric field $\overline{D}- = (1/\mu)\,(i\,E\,\sin\theta) = (1/\mu)\,(\overline{E}\,\sin\theta)$, we now have the involvement of a "capacitive-displacement" version of the electric field, which may be denoted by the symbol $D+$ (or alternatively by $E_{\mathrm{d}}+$), where the "+" sign indicates the capacitive nature of the displacement. The relation describing this field version is $D+=\varepsilon E$, where ε is the electric permittivity, a measure of electric polarizability of dielectrics. As mentioned earlier, the role of ε here is far more extensive than only being a measure of polarizability, as its units come in Farad per meter, where Farad is the unit of capacitance. Hence, the existence of ε is actually what allows (and accounts) for the capacitive shift in *time* of the Current oscillation with respect to the Voltage oscillation, where that has a reverse analogy to a displacement-drag of Voltage with respect to Current. (Note that later we shall figure that an analogous shift in *space* accounts for the effect of barrier penetration, via a symmetric way). Capacitance may be a complex entity (involving a real as well as an imaginary part), and because of that the displacement electric field gets to be more accurately described through the following relation.

$$D+ = \varepsilon \, (E \cos\theta - i \, E \sin\theta) \qquad\qquad (6\text{-}30)$$

The fact that $D+$ has two coordinates (a cosine and a sine one) allows it to also become involved in two orthogonal interactions. Here, we are concerned with the second portion of (6-30) which corresponds to

$$\bar{D}+ = \ \varepsilon \, (-i \, E \sin\theta) = - \, \varepsilon \, (\bar{E} \sin\theta) \qquad\qquad (6\text{-}31)$$

This coordinate concerns an imaginary (negative energy) version of the electric field (where the negative energy relates to the existence of the term i, and not to the minus sign). Under certain conditions this imaginary field version may produce real outcome, for instance if it interacts with the imaginary part of a particle's wave nature, so the two imaginary terms produce a negative real result. (This negative real result may actually combine with the negative sign to yield a positive outcome). The displacement shift affects on the particle's wave function and brings the equivalent of a displacement current I_d that develops as per the charge's wave nature, and concerns the **displacement of current between the plates of a capacitor**. Actually, as the interaction involves the imaginary part of the particle's wave function, it should be more accurate to consider the displacement current to concern transfer of **anticharge** (which was presented earlier to associate the imaginary part of electron's wave nature). The displacement of anticharge is mediated by the alternation of the imaginary electric field coordinate, and as that is a wave effect, it resembles the effect of frustrated total internal reflection, here developing in *time*. (This actually seems to also be symmetric to a displacement in *space*, which we shall later explain to concern the tunneling of matter in the effect of barrier penetration). Because of the imaginary essence of the interaction, the corresponding displacement field becomes evanescent in short range over the real world, thus the feasibility in displacing charge becomes significantly reduced as the width between the capacitor's plates increases.

As the displacement concerns a shift in time of charges' wave nature, the way it works has nothing to do with conventional motion of charges as per their particle nature, even though these two processes may illusively be considered very similar. In fact, during the phase shift the displaced portion of electrons should not be producing a real magnetic field. It should instead produce an imaginary version of the magnetic field, which refers to a magnetomotive potential as considered above.

Before we leave this section, we shall apply a modification in the way the above equations are expressed, for the sake of symmetry with other effects of reaction. Referring to $\bar{B}$ (loosely referring to $\oint B$) and making use of an earlier assumption that the units of voltage may resemble dt/dx, we have

$$(dx/L)\, \bar{B}\; dx = C\, V/\Delta t$$
$$(dx/\Delta t)\, \bar{B}\; dx = C\, (dt/\Delta x)/\Delta t$$
$$(dx/\Delta t)\, \bar{B} = \cancel{(dt/\Delta t)}\; C\, (1/dx^2)$$
$$(1/\Delta t)\, \bar{B} = [C\, (1/dx^2)]\, (1/dx)$$
$$\bar{B} / \Delta t = \Phi_C/dx \qquad (6\text{-}32)$$

In the above calculation we have assumed a new concept, of the quantity $C(1/dx^2)$ to correspond to a "flux of capacitance" which resembles a displacement of charge as shown in (6-33). Actually, since the denominator in the term C/dx^2 concerns ordinary space (and not reciprocal space $C/(1/dx^2)$), it could be named "flow" of capacitance, instead of flux.

$$\Phi_C = C\, (1/dx^2) = C/dx^2 \qquad (6\text{-}33)$$

As per the above, the transitional effect that takes place in a capacitor may be thought of to correspond to a **"displacement capacitive"** effect (accounted for as a "capacitive flow" of charge per space), as the right side of (6-32) describes. That in turn, goes along with the antimagnetic field (the imaginary coordinate of the magnetic field) which reacts with respect to the shift in time Δt, where that corresponds to the **magnetomotive potential**, accounted for in the left side of that relation. The field lines of that potential are effectively radial-like (open ended, while they may be considered to follow a hypothetical loop through the very distant universe), and correspond to an anti-magnetic field which is orthogonal to the conventional magnetic field lines. Furthermore, the relation (6-32) could also be re-arranged so as to involve the "displacement antimagnetic" field, as follows:

$$(dx/\Delta t)\, \bar{B} = \Phi_C$$
$$(dx/L)\, \bar{B} = \Phi_C$$
$$(1/\mu)\, \bar{B} = \Phi_C$$
$$\bar{H} = \Phi_C \qquad (6\text{-}34)$$

This representation is rather looser but handy in terms of comparison with other fields considered (like in (6-7) or (6-18)). What it may tell for the case of the capacitor, is that the imaginary coordinate of the displacement magnetic field applies as a capacitive flow (which concerns the displacement current).

Notice that the capacitive phase shift gets to borrow energy from the time domain, and that applies as a "backing" (counterbalancing) potential toward the electric source of the circuit. Thus, it constitutes a way of conserving the **electronic counterpart of momentum**. It is of significance that such backing applies almost instantaneously throughout the circuit loop, as it is not feasible to "borrow an opposite force in order to apply it later" (since in such case there wouldn't be any "backing" for exerting the original force). This holds even for interactions of quantum nature, where, if an opposite force does not concurrently apply directly upon an end-surface, it gets to be delivered to a mediator particle of appropriate effective charge and mass, as it holds for the electroweak interaction's bosons which we shall consider later.

On a general side note that concerns all of the above, even though capacitance C has been associated to a phase shift forward in time $+\Delta t$ and inductance L has been associated to a phase shift backward in time $-\Delta t$, it does NOT hold $C=-L$. The reason is that either of the two shifts develops in opposite sign of energy with respect to the other, thus either one is imaginary with respect to the other. This is why, in typical equations involving both C and L, one of the two needs to be accounted for as a negative reciprocal, like for instance involving L in the one side and $1/C$ in the other side (or vice versa) for the equality to be valid in the real world.

(b) Reaction causing a conduction current (in an electric battery)

According to conventional theory the operation of the electric battery is based in the electromotive force. However, this statement seems not accurate since the electromotive force causes either *impedance* (in reacting a change in the motional status of charges through the effect of self-induction) or causes a reactive *displacement* current (affected on the imaginary portion of electrons', like in the case of a temporal current in a metal ring when there is a change in magnetic *flux* through the ring).

The magnetomotive potential, instead, seems possible to operate in a complementary way that allows for an enhancing interaction to take place (instead of an impeding one). This is possible through the interaction of two different imaginary entities that bring a negative-real outcome, and then, along with a negation, that outcome may turn into positive. In this way we may get energy flow from the imaginary domain toward the real domain. The corresponding enhancing action could be named by using the prefix "push-" (instead of "back-" which was used for the back-electromotive force). For reasons of symmetry in the naming, the corresponding reaction potential could be called a **"push-magnetomotive potential"**. To figure a relation that would describe the way this works, we may start with a reference to -probably the closest- available relation on this subject, which corresponds to Ampere's law:

$$1/\mu_o \oint \boldsymbol{B} \cdot d\ell = I \qquad\qquad \text{(5-3) or (6-35)}$$

This relation describes the magnetic field across a current-carrying loop, so it is not really too close to what we are looking for, but we could at least rely on it so as to device another equation that would be closer to describing the operation of the electric battery. For doing that, we could modify the relation (6-35), so as to have it become symmetric to the expression of the electromotive force $\oint \boldsymbol{E} \cdot d\ell = -L\, dI/dt$. Such symmetry is reached in (6-36) :

$$\oint \boldsymbol{B} \cdot d\ell = \mu_o\, I \quad \rightarrow$$
$$\oint \boldsymbol{B} \cdot d\ell = (L/dx)\, I \quad \rightarrow$$
$$\oint \boldsymbol{B} \cdot d\ell = L\, dI/dx \qquad\qquad \text{(6-36)}$$

A point of interest is that the right side of this expression describes a change in electric current *per space* (instead of *per time*). In addition, this seems like an inductive type of relation, without any obvious involvement of a *voltage* term which would be in need in order to account for the voltage that an electric battery exhibits. We should therefore try a modified approximation.

The modified approximation could make use of two known relations: That for the electric current it holds $I=Q/t$, and for the charge it holds $Q=CV$. Actually, in relation to material to follow in section 8, it arises that the relation $Q=CV$ is valid even at the elementary particle level, concerning electron's charge (which is quantized much like the energy associated to the particle's matter is quantized as per the Planck constant). By plugging-in these two relations to Ampere's law, we come up to the relation

$$\oint \boldsymbol{B} \cdot d\ell = C\, V/dx \qquad\qquad \text{(6-37)}$$

The latter involves a voltage term, so it is much closer to what we are searching for. It looks much like (6-25), but now involves *conventional* voltage V instead of *displacement* voltage V_d. The difference between these two, is that the displacement voltage seems to relate to an *imaginary* coordinate of the capacitive displacement magnetic field (6-28), while the conventional voltage seems to relate to a *real* coordinate of the capacitive displacement magnetic field, which could have as follows

$$H+= (1/\varepsilon)\ \oint B = (1/\varepsilon)\ (B\ \cos\theta) \tag{6-38}$$

The relations (6-28) and (6-38) describe fields which are involved in interactions mediating opposite flow of energy, the former seemingly associating to the displacement current through a capacitor, and the latter seemingly associating to a conduction current generated by an electric battery. The underlying difference between these two types of current appears to have a deeper root as follows:

- A *displacement* current (like the one developing between the plates of a capacitor) is generated by an alternation of polarity of the plates. However, that concerns an alternation in the number of electrons in each plate, where all electrons carry negative charge.

- A conduction current (like the one developing by an electric battery), is generated through a completely different way, as the alternation here seems to involve the wave nature of elementary particles, and more particularly it seems to make use of a specific characteristic of composite particles, which traces within their constituent quarks. As it will be described in section 10, quarks behave as leptons that are subject to an internal phase shift which gets a portion of each quark to dive (reside) in opposite energy. This allows for negative energy characteristics to interact in yielding an outcome in positive energy. As we shall figure, quarks may reach stable form when combining with each other, in which case they may exhibit stable *positive* charge (while instead, positive charge would be unstable, just like the positively charged positrons are not stable, as they quickly annihilate). That portions of each quark reside in opposite energy, also provides the necessary negation required for the total interaction that powers a battery to have an enhancing instead of an impeding character. The interaction provides an energy surplus through a similar reasoning to as it was explained for the cosmic region b'-d to yield positive energy, in figure 4-2.

In this sense, the cosine coordinate in (6-38) represents the portion of energy that projects toward the real world and produces the enhancing outcome. The angle θ in this case does NOT concern a phase shift between voltage and current (as it applies in electronics). Instead, it refers to an internal property of quarks which concern the Weinberg angle (to be considered in section 10).

On top of the above, in the next two sections we shall also figure that elementary particles' quantum nature involves a type of *oscillation* in capacitance, where that oscillation is counterbalanced by the existence of the particle's *steady* electric field. Notice that the condition of an *alternating* capacitance and a *steady* electric field in a quantum particle, is inverse to what holds in electric circuits oscillations, where capacitance is *steady* and it's the electric field that *oscillates*. (This inversion will later be shown to constitute the grounds of the difference in behavior between fermions and bosons).

Likewise, the *steady* magnetomotive potential that becomes involved in powering the electric battery seems to arise in respond to a *cyclic* behavior of capacitance. In quantitative terms, this applies much like a different way of reading the right side of (6-24), where, the electric potential V stays fixed, and it's the capacitance C that oscillates. In other words, that condition corresponds to "reading" that side of the equation as an oscillation of $[V\,(dC/dt)]$ instead of $[C\,(dV/dt)]$. And that is counterbalanced by the capacitive displacement magnetomotive potential $1/\mu_o \oint \boldsymbol{B} \cdot d\ell$, that the left side of the equation describes.

This non-conventional effect of reaction seems to bring an enhancing outcome (instead of an impeding one) and we may call that **self-capacitance** (by symmetry in the naming to the effect of self-induction). And while self-induction was associated to a "lag in current", the self-capacitance can be considered to concern a "lag in voltage", where that resembles a "lead in current". When that applies with positive charge (which relates to opposite energy in time) it enhances the flow of energy (instead of impeding it). Thus, it seems that the interaction that takes place within an electric battery makes use of that particular nature of quarks (which we shall describe in section 10), and gets them interact collectively as if they partially contained antimatter (hence their stable positive charge).

Notice that, since antimatter rides (lives in) negative flow of time, its essence is not imaginary with respect to the imaginary version of the magnetic field which drives the magnetomotive potential. Hence, the potential $\oint \boldsymbol{B} \cdot d\ell$ may apply on antimatter's real portion (as being particle-like) not its imaginary portion (described through its wave nature). And since the interaction involves particle-like nature, the potential should generate *conduction* current of positive charge (instead of *displacement* current). The corresponding field version involved in this case seems to be the one described in (6-38). The equation which describes the transfer of positive charge may conceptually be as follows, and that should reflect what holds inside the battery.

$$(1/\mu_o)\oint \boldsymbol{B}\cdot d\ell = [V\,(dC/dt)]$$
$$(dx/\text{-}\Delta t)\oint \boldsymbol{B}\cdot d\ell = [V\,(dC/dt)]$$
$$\oint \boldsymbol{B}\cdot d\ell = (\Delta t/dx)\,[V\,(dC/dt)]$$
$$\oint \boldsymbol{B}\cdot d\ell = [V\,(dC/dx)] \qquad\qquad (6\text{-}39)$$

Mind the difference that, in self-capacitance (6-39) there is change in capacitance with respect to space (dx), unlike self-inductance ($\oint \boldsymbol{E}\cdot d\ell = \text{-}L\,dI/dt$) where there is a change in current taking place with respect to time (dt). Since the "oscillation" of capacitance deploys on space (in the form of a "loop-like waviness"), if we transfer the term (dC/dx) to the left side of the equation it takes the form of reciprocal permittivity

$$(dx/\Delta t)\oint \boldsymbol{B}\cdot d\ell = V$$
$$(1/\varepsilon_o)\oint \boldsymbol{B}\cdot d\ell = V$$
$$(1/\varepsilon_o)\oint \boldsymbol{B} = V\,/\,d\ell \qquad\qquad (6\text{-}40)$$

It seems that this potential described in the left side of this relation is an electric-like potential, and has similar nature to the electric field and potential of the positively charged protons. It describes an interaction of enhancing character, in contrast to (6-2) of self-induction which reads $-(1/\mu)\oint \boldsymbol{E} = dI/dt$ and has impeding character. Actually, equation (6-40) could be written in the inverse direction, since the voltage V that arises is not the "cause" but is rather the "effect" of the process. That is,

$$V = (1/\varepsilon_o)\oint \boldsymbol{B}\cdot d\ell \qquad\qquad (6\text{-}41)$$

As this enhancing interaction may develop collectively by neighboring particles at the atomic scale, it takes the form of a chemical tendency in applying a positive magnetomotive potential. Since this potential is radial-like, and the units of $(1/\varepsilon_o)\oint \boldsymbol{B}$ resemble the units of $vB = E$ as explained previously, the way it applies resembles the way of action of the electric field. This process seems to address the process that happens internally to the electric battery and creates the potential difference between its poles. While in turn, that tends to drive a parallel flow of negative charge through a circuit outside the battery.

Looking into the technicals of a physical battery, according to conventional knowledge the battery's anode interacts with an electrolyte and loses electrons, creating positive ions (the process called *oxidation*). It seems that during the oxidation electrons are being detached via an enhancing "capacitive displacement magnetic" potential (the magnetomotive potential), which is an effect of capacitive nature. The ions tend to reach a stable state so they move through the battery's electrolyte in order to reach the cathode and re-unite with electrons coming through the external circuit (the process called *reduction*).

The opposite holds when reduction takes place. The total process keeps being fed as the interaction keeps going on collectively inside the battery, where the **magnetomotive potential** provides higher energy in the anode than in the cathode, and that is what powers the creation of a conduction current between its poles (externally to the battery).

Note that a symmetric process takes place in mechanics and is responsible for the powering of biochemical reactions which support life, as will be considered next.

6.4 The Gravitomagnetomotive Potential

The gravitomagnetomotive interaction involves an exchange in the parameters of space and time with respect to what holds for the gravitomotive interaction. This swapping of parameters additionally requires the involvement of the mechano-permittivity and mechano-permeability terms. It appears that the particular units of these terms $\varepsilon_g=+\Delta x/dt$ and $(1/\mu_g)=dt/-\Delta x$ are such, that their involvement applies as an indirect way of turning back to the alternations taking place with respect to the parameters of space and time that hold in the gravitomotive case. Furthermore, the gravitomagnetomotive effect exhibits in two basic ways too, one involving a change in field flux, and another involving a change in the flow of particles.

(a) Reaction to a change in gravitomagnetic flux (Barrier Penetration)
In analogy to as the magnetomotive effect allows for the transfer of a displacement current through a capacitor's plates, a very similar effect takes place in mechanics in the case of barrier penetration, where a particle's wave (which shall be described to concern an oscillation in the metric of time) tunnels (in space) through a potential barrier. While this is treated as a quantum mechanical process, it is a wave effect that could be approximated in classical physics terms as an effect involving the "**gravitomagnetomotive**" potential which causes "displacement momentum" through the barrier (much like the displacement current develops through the gap between the plates of a capacitor). The displacement momentum, denoted as p_d should be lasting for as long as the matter wave oscillation takes place, in analogy to as it holds with

the displacement current I_d in electronics, whose displacement lasts for as long as the electric flux oscillates.

In quantitative terms the gravitomagnetomotive potential may be described through a mechanical counterpart of Maxwell's fourth law, which may take the form of (6-42). The involvement of mechano-permittivity and mechano-permeability makes the interaction subtler, since the $\varepsilon_g=C_g/dt$ involves mechano-capacitance C_g (measured in mechano-Farad per second) and $\mu_g=L_g/dt$ involves mechano-inductance L_g (measured in mechano-Henries per second), where both ε_g and μ_g are typically complex numbers. The corresponding interaction may be described as follows

$$1/\mu_g \oint \boldsymbol{T} \cdot d\ell_t = \varepsilon_g \, \theta\Phi_g/\theta x \qquad\qquad \text{(5-40) or (6-42)}$$

Substitution of the terms ε_g and μ_g allows to re-write (6-42) in the form (6-43), which involves the displacement field versions $T_d=(1/\mu_g)T$ and $g_d=\varepsilon_g g$.

$$\oint \boldsymbol{T_d} \cdot d\ell_t = \theta\Phi_{g_d}/\theta x \qquad\qquad \text{(6-43)}$$

Since mechano-capacitance and mechano-inductance may be complex entities, that requires for the displacement-gravitational and the displacement-gravitomagnetic fields to be complex entities as well. In such case it should hold $g_d=\varepsilon_g(g \cos\theta - ig \sin\theta)$ and $T_d=(1/\mu_g)(-T\cos\theta + iT\sin\theta)$ respectively, where θ corresponds to the Weinberg angle. In this sense the gravitomagnetomotive effect can be considered to materialize in two ways which are orthogonal to each other. In the one, **the *imaginary* coordinate seems to account for the displacement of matter in the form of "displacement momentum"** p_d, while the *real* coordinate seems to concern the involvement of the gravitomagnetic field in interactions in positive energy.

An example of displacement momentum (referring to the term $\varepsilon_g\theta\Phi_g/\theta x=p_d$) via the gravitomagnetomotive effect concerns the effect of barrier penetration, which materializes in a symmetric way to as current is displaced between the plates of a capacitor in electronics (referring to the term $\varepsilon_o\theta\Phi_E/\theta t=I_d$). The subscript $_d$ next to the momentum denotes the displacement nature of momentum transfer. The displacement momentum has the same units with the conventional momentum, but it is only temporal and lasts for as long at the waviness of the incident matter wave lump lasts, where **this waviness applies as an alternating gravitational flux**. And while barrier penetration is a wave effect, the displacement of a portion of the matter wave through the barrier takes place like an effect of **frustrated total internal reflection** (of the matter wave). Here, the displacement associates to a mechano-capacitive shift *in* *space*. The coordinate of the wave which makes it through the barrier carries imaginary characteristics and therefore behaves as an **evanescent wave**. The imaginary nature of the interaction implies that it should bear the characteristics of a **tunneling** process, in particular a tunneling in space.

In classical terms one would expect for the corresponding displacement to carry energy through the barrier region, however this notion is misleading since there should be no conventional energy flow through the barrier. Instead of "flow" (which would ascribe to a real entity), we have a "flux" (since it concerns the displacement of an imaginary entity). In such case, the imaginary component may be understood to displace "**negative momentum**" (which is not a real entity). In terms of energy involved, the flux involved seems to actually correspond to the following, where Φg_d corresponds to displacement gravitational flux, and θ corresponds to the Weinberg angle.

$$\varepsilon_g \, (\theta \Phi g \, sin\theta)/dx = (\theta \Phi g_d \, sin\theta)/dx \qquad (6\text{-}44)$$

Speaking of gravitational flux Φ_g, we should consider that it relates to the gravitational field g in a similar way to as the electric flux Φ_E relates to the electric field E. And, just like an alternation in electric flux generates a displacement current, an alternation in gravitational flux generates displacement momentum. **An alternation in Φ_g seems to actually concern the alternation in the particle's wave function**, which will later be associated to an oscillation in the metric of time. In relation to that, the gravitational flux Φ_g should be represented by a relation as follows, where that arises in symmetry to its electric flux counterpart $\Phi_E = E \, dx^2$.

$$\Phi_g = g \, dt^2 = g \, / \, (1/dt^2) \qquad (6\text{-}45)$$

That the term $(1/dt^2)$ involves the power of two may be assumed to consider the existence of two (out of a total of three) dimensions of time, instead of only one dimension as conventionally accepted, but this should be fine as explained earlier during section 6.1.

Other than the relation (6-45), the gravitational flux may be alternatively represented by a relation like (6-46), where that again comes in symmetry to its electric counterpart being $\Phi_E = Q/\varepsilon_o = (1/\varepsilon_o) I dt$.

$$\Phi_g = h/\varepsilon_g = (1/\varepsilon_g) \, p \, dx \qquad (6\text{-}46)$$

Since for the gravitational flux it holds $\Phi_g = h/\varepsilon_g$, the right side of (6-42) may turn to $\varepsilon_g \, \theta\Phi_g/\theta x \rightarrow \theta h/\theta x$, where that refers to the momentum $p = h/\lambda$. In that sense, the relation (6-42) seems to reduce to the equation (6-47) which looks symmetric to (6-23) of electromagnetism.

$$(1/\mu_\mathrm{g})\oint T\cdot dA = p_\mathrm{d} \tag{6-47}$$

The right side of (6-47) arose through the use of (6-46) which contains mechano-capacitance C_g (within the term ε_g), where mechano-capacitance concerns a phase shift in space $+\Delta x$, so this gets the right side of (6-47) to describe "displacement momentum" p_d (instead of a conventional momentum p). Actually, the degree of displacement taking places in this case relates to the shape, size, contents, and extend of the barrier, which defines the value of the corresponding mechano-capacitance $C_\mathrm{g}=+\Delta x$.

The relation (6-47) may allow to explain in intuitive terms the way that the mechano-capacitive shift may take place in the space domain, through a similar pattern as the displacement of charge was explained to take place between the plates of a capacitor. There, we relied in the equations $I=Q/dt$ and $Q=CV$, and assumed that these equations should apply for the displacement current as well, so we were able to obtain the relation (6-24). Here, we may use the relation $p=h/\lambda=h/dx$ (instead of $I=Q/dt$), however we need to device a relation that would account for the Planck constant h (in symmetry to the relation $Q=CV$). For doing this, we may follow the following conceptual pattern

$$\begin{aligned}
p &= h/\lambda = mv \\
h &= mv\lambda \\
h &= C_\mathrm{g}\,v
\end{aligned} \tag{6-48}$$

The symmetry between (6-48) and the electric counterpart $Q=CV$ may be noticed via a reference to the involved units. The capacitance C concerns a shift in time Δt, while the mechano-capacitance C_g concerns a shift in space, so they differ by a swapping between time and space. Likewise, the voltage V was previously explained to associate to units which resemble the reciprocal of velocity dt/dx, so here again V and v differ by a swapping of the parameters of time and space. However, in an initial reading the velocity $v=dx/dt$ does not seem to reflect a potential (in a similar sense to as "voltage" corresponds to a potential). In such case, it turns out that **when it comes to an inverse spacetime curvature, where space and time swap roles, the classical velocity applies as a potential (that is, as a pressure) from the inversed reference point of view**. This explains the symmetry between the relation describing the velocity $v=g\,dt$ (where g is the gravitational field), and the relation for voltage $V=E\,dx$ where E is the electric field.

Furthermore, for (6-48) to hold, we have assumed that mechano-capacitance corresponds to

$C_g = m\lambda$ which may also be represented as

$$C_g = m / (1/dx) \tag{6-49}$$

Here, we are adopting classical physics terms in order to describe a quantum effect. That is, we are involving a particle-like notion (of mass) in an effort to describe a wave-like notion which concern a shift of the matter wave in the metric of space. For this, we shall grant to ourselves a bit of freedom in terms of formulation, in order to be able to approximate the matching, and describe a quantum effect in a classical-physics-like way, which involves a shift in space. In accordance to that, it seems that the mechano-capacitance C_g corresponds to a shift in the space metric Δx, and this translates to a local shift in the wavelength λ of a particle. The shift in space can be accounted for as an imaginary effect, and as explained earlier an imaginary effect may project to the real world through its negative reciprocal (since it holds $i^2=-1 \rightarrow i=-1/i$). Therefore, this may be represented as a shift of the particle matter wave with respect to reciprocal space ($1/dx$), which is what (6-49) loosely describes. Following the above, (6-42) may take a form as follows:

$$1/\mu_g \oint T \cdot d\ell_t = C_g\, v_d/\Delta\lambda \tag{6-50}$$

Notice how symmetric this is to (6-24) of the magnetomotive case, which reads $1/\mu_o \oint B \cdot d\ell = C\, V_d/dt$. Furthermore, if we transfer the mechano-permeability term $\mu_g=(-\Delta x/dt)$ to the right side, that becomes $(-\Delta x/dt)\, C_g\, v/\Delta\lambda$, where, by canceling out the numerator Δx and the denominator $\Delta\lambda$ (as they both concern units of space), we get

$$\oint T \cdot d\ell_t = C_g\, v_d/dt \tag{6-51}$$

Notice here a new symmetry, between this expression which describes the gravito-magnetomotive potential, and its electromotive counterpart (6-1) which reads $\oint E \cdot d\ell = -L\, dI/dt$. In fact, this comparison is significant as it hints that the right side of (6-1) could have also involved a similar cancelation, therefore it should have rather looked like $(-\Delta x/dt)\, L\, dI/dx$, where that refers to $(\mu_g)\, L\, dI/dx$, and by moving the mechano-permeability to the left side, the relation (6-1) should take the form (6-26) which reads $(1/\mu_g) \oint E \cdot d\ell = -L\, dI/dx$ and seems more complete. (Notice that despite the fact that (6-1) is an equation from the electromagnetic interaction, here it uses the mechano-permeability term from the gravitomagnetic interaction).

Likewise, it should similarly hold for that equation's gravitational counterpart (6-11) which reads $\oint g \, d\ell_t = -L_g \, dp/dx$. That may have also gone through a similar cancelation, in which case its right side could be $(\Delta t/dx) -L_g \, dp/dt$, where that refers to $(\mu_o) -L_g \, dp/dt$. And, by moving the mechano-permeability to the left side, the relation (6-11) should have taken the following form which seems to be more complete (again, despite the fact that (6-11) refers to the gravitomagnetic interaction but uses the permeability term from electromagnetism).

$$(1/\mu_o) \oint g \, d\ell_t = -L_g \, dp/dt \qquad (6\text{-}52)$$

We shall now revert to (6-51) and go a step further, transferring the mechano-capacitance C_g from the right to the left side of it

$$(1/C_g) \oint \boldsymbol{T} \cdot d\ell_t = v_d/dt$$
$$(d\ell_t/C_g) \oint \boldsymbol{T} = v_d/dt$$
$$(1/\varepsilon_g) \oint \boldsymbol{T} = v_d/dt \qquad (6\text{-}53)$$

Here, the existence of the mechano-permittivity ε_g (instead of mechano-permeability μ_g) next to the imaginary version of the magnetic field $\oint \boldsymbol{T}$ is not by coincidence. Since the interaction applies in the imaginary domain, it concerns the viewpoint of negative energy, thus the shift in space relating to the gravitomagnetic field is mechano-capacitive instead of mechano-inductive. That we have ε_g next to the gravitomagnetic field (instead of μ_g) is also fine in the sense that both these terms have same units, they only differ in the sign of the involved shift in space, now being $+\Delta x$ instead of $-\Delta x$. Actually, since the interaction involves the point of view of negative energy, there is no need for a reciprocation from $(1/\varepsilon_g)$ to (ε_g). What (6-53) may tell, is that a change in velocity (where velocity should be perceived as a voltage-like potential from the viewpoint of opposite energy) is reacted by a capacitive **displacement** version of the gravitomagnetic field, where the shift in space effectively concerns a tunneling effect across the barrier.

Notice the symmetry between (6-53) and (6-27) of the magnetomotive case which reads $(1/\varepsilon_o)\oint \boldsymbol{B}=V_d/dx$. In the gravitomagnetomotive case, the capacitive displacement imaginary version of the gravitomagnetic field may be referred to as follows

$$\bar{T}_d += (1/\varepsilon_g) \, (- i \, T \sin\varphi) = (1/\varepsilon_g) \, (-\bar{T} \, \sin\theta) \qquad (6\text{-}54)$$

Here, the term T_d indicates a "displacement gravitomagnetic field", the upper bar indicates the imaginary version of the field (therefore reflecting a displacement "negative energy field"), while the plus sign at its right indicates its "capacitive" character, which differentiates it from the "inductive" displacement magnetic field $T_d=(1/\mu_g)T$ which is responsible for the deflection of a moving matter. Notice the symmetry between (6-54) and (6-28) of the magnetomotive case. While this field is a reaction field, it should involve opposite energy, and should therefore be accounted for as a projection toward the real axis, where it leaks energy toward. In terms of energy this projection could be accompanied by a **negative sine** coordinate of an angle θ, where θ identifies to the phase shift between the parameters of Space and Time, where that may be influenced by the shape and size of the barrier (just like a capacitive phase shift is influenced by the shape of the capacitor's plates in electronics). The sine coordinate concerns the energy portion retrieved (and consumed) from the amplitude of the matter wave without doing work, so it applies as an impeding agent that reacts the transmission through the barrier.

If we allow ourselves to transfer the term dt from the right to the left side of (6-53) and assume that dt resembles $d\ell$, then we may get an Ampere's-law-like expression which concerns how the imaginary version of the displacement gravitomagnetic field relates to the displacement of particles' wave function by a displacement velocity v_d (instead of conventional motion v).

$$(1/\varepsilon_g) \oint T = v_d/dt$$
$$(1/\varepsilon_g) \oint T \cdot dt = v_d$$
$$(1/\varepsilon_g) \oint T \cdot d\ell_t = v_d \tag{6-55}$$

Coming back to (6-53), one may notice that such relation should rather be written in reverse, since the displacement velocity (which physically has the form of a pressure gradient as perceived from the negative energy point of view where velocity applies as a potential) is not the "cause" but it is rather the "effect" of the process. (That is symmetric to what was also described in relation to the expression (6-29) of the magnetomotive case). Furthermore, the displacement *lag* in velocity (having the form of a potential) with respect to momentum (as per the wave nature $p=h/\lambda$) may alternatively be treated as a *lead* in momentum with respect to the potential, which is what mechano-capacitance $+\Delta x$ could be understood to associate with.

To consider the role of the change in gravitational flux $\varepsilon_g \partial\Phi_g/\partial x$ in the interaction, is like approximating the same interaction from the gravitational field's point of view. Similar to as in electronics the displacement current I_d between the capacitor's plates arises from a change in electric flux $\varepsilon\partial\Phi_E/\partial t$

which in turn creates a displacement potential $(1/\varepsilon)\oint B$, in mechanics we need to figure how the displacement potential term $(1/\varepsilon_g)\oint T$ arises due to a change in displacement gravitational flux per space $\varepsilon_g\,\partial\Phi_g/\partial x$ which equals p_d.

The term $\varepsilon_g\,\partial\Phi_g$ derives from the displacement gravitational field, and one may recall that in (6-14) we considered the "inductive-displacement" version of that field $\bar{g}_{d-} = (1/\mu_g)\oint g = -(1/\mu_g)(i\,g\,\sin\theta) = -(1/\mu_g)(\bar{g}\,\sin\theta)$. Now instead, we have the involvement of a "capacitive-displacement" version of the gravitational field, which may be denoted by the symbol $\bar{g}_d{+}$, where the "+" sign indicates the capacitive nature of the displacement. This field corresponds to $\bar{g}_d{+} = \varepsilon_g\,g$ (in symmetry to the field $D{+}{=}\varepsilon E$ of the electric case). The mechano-capacitance ε_g is typically a complex entity (including a real as well as an imaginary part), and because of that the displacement gravitational field gets to be more accurately described through the relation (6-56). The two coordinates allow it to become involved in *two* orthogonal interactions, and we are here concerned with the *one* referring to the sine coordinate (6-57).

$$\bar{g}_d{+} = \varepsilon_g\,(g\,\cos\theta - i\,g\,\sin\theta) \tag{6-56}$$

$$\bar{g}_d{+}_{(im)} = -\,\varepsilon_g\,(i\,g\,\sin\theta) = -\,\varepsilon_g\,(\bar{g}\,\sin\theta) \tag{6-57}$$

This coordinate concerns an imaginary (negative energy) version of the displacement gravitational field, which may produce real outcome when interacting with the imaginary part of a particle's wave function, so the two imaginary terms produce a negative real result. Actually, the *negative* real result may combine with the reaction nature of the interaction, and yield a positive result. The displacement character of this field associates to a shift forward in *space* of the particle's wave function, which brings the equivalent of **displacing momentum in barrier penetration**. The corresponding displacement of this imaginary field coordinate in *space* functions like the effect of frustrated total internal reflection. (This is also symmetric to a displacement in *time* which was explained to concern tunneling of charge between the plates of a capacitor). Because of the imaginary essence of the interaction, the corresponding displacement field becomes evanescent in short range over the real world, thus the feasibility of displacement of the matter wave becomes significantly reduced as the width of the barrier increases. Furthermore, as the effect of displacement is an effect of wave nature, the way it works has nothing to do with conventional motion of matter as per the particle nature. In line to the above, Table 6.2 provides side-by-side a comparison of the quantitative symmetry between the displacement of momentum in barrier penetration, and the displacement of current through the plates of a capacitor.

Gravitomagnetomotive case	Magnetomotive case
$1/\mu_g \oint \boldsymbol{T} \cdot d\ell_t = \varepsilon_g\, \theta\Phi_g/\theta x$	$1/\mu_o \oint \boldsymbol{B} \cdot d\ell = \varepsilon_o\, \theta\Phi_E/\theta t$
and since $\Phi_g=(h/\varepsilon_g)=(1/\varepsilon_g)pdx$, we get	and since $\Phi_E=Q/\varepsilon_o=(1/\varepsilon_o)Idt$, we get
$1/\mu_g \oint \boldsymbol{T} \cdot d\ell_t = p_d$	$1/\mu_o \oint \boldsymbol{B}{\cdot}d\ell = I_d$
and since $p_d=h_d/\Delta x$ and $h=mv\lambda=C_g v$	and since $I=Q/t$ and $Q=CV$ we get
$1/\mu_g \oint \boldsymbol{T} \cdot d\ell_t = C_g\, v/\Delta\lambda$	$1/\mu_o \oint \boldsymbol{B}{\cdot}d\ell = C\, V/dt$
$(dt/L_g)\,\bar{T}\,dt = C_g\, v\,/\,dx$	$(dx/L)\,\bar{B}\,dx = C\, V/\Delta t$
$(dt/\Delta x)\,\bar{T}\,dt=C_g\,(dx/dt)/dx$	$(dx/\Delta t)\,\bar{B}\,dx=C\,(dt/dx)/\Delta t$
$(dt/\Delta x)\,\bar{T} = C_g\,(1/dt^2)$	$(dx/\Delta t)\,\bar{B} = C\,(1/dx^2)$
$(1/\Delta x)\,\bar{T} = [C_g(1/dt^2)](1/dt)$	$(1/\Delta t)\,\bar{B} = [C(1/dx^2)](1/dx)$
$\bar{T}/\,\Delta x = \Phi_{Cg}/dt$	$\bar{B}\,/\,\Delta t = \Phi_C/dx$
or	or
$(dt/\Delta x)\,\bar{T} = \Phi_{Cg}$	$(dx/\Delta t)\,\bar{B} = \Phi_C$
$(dt/L_g)\,\bar{T} = \Phi_{Cg}$	$(dx/L)\,\bar{B} = \Phi_C$
$(1/\mu_g)\,\bar{T} = \Phi_{Cg}$	$(1/\mu)\,\bar{B} = \Phi_C$
$\bar{T}_d = \Phi_{Cg}$	$\bar{H} = \Phi_C$

Table 6.2: Magnetomotive vs. Gravitomagnetomotive relations

In the above calculations we have assumed a new concept, of the "flux of mechano-capacitance" $\Phi_{Cg}=C_g(1/dt^2)$. This resembles a displacement of momentum (in analogy to as the "flux of capacitance" in electronics resembles a displacement of charge). The calculations arrive to the relation

$$\bar{T}/\Delta x = \Phi_{Cg}/dt \qquad\qquad (6\text{-}58)$$

In this sense, the transitional effect that takes place in barrier penetration corresponds to **"displacement mechano-capacitive"** effect (accounted for as a change in "mechano-capacitive flux" per time), as the right side of (6-58) describes. That in turn, goes along with the negative gravitomagnetic field (the imaginary coordinate of the gravitomagnetic field) changing with respect to the shift in space Δx, creating the **gravitomagnetomotive potential** (a sort of negative pressure gradient) which is reflected in the left side of this relation. The field lines of the corresponding potential are effectively radial-like (they appear to be open ended, however they should be considered to follow a hypothetical loop through the distant universe), and correspond to a negative gravitomagnetic field which is orthogonal to the "conventional" gravitomagnetic field lines.

Notice how **the symmetry between the terms ($\varepsilon_g \ \theta\Phi_g/\theta x$) and ($\varepsilon_o \ \theta\Phi_E/\theta t$) in table 6.2 allows to provide meaningful essence to the oscillating quantity Ψ in a particle's wave function, giving to it an "oscillating displacement gravitational flux" analogy, in symmetry to the "oscillating displacement electric flux" in electronics**. Furthermore, during barrier penetration we have a negative energy effect taking place, where the matter wave's alternation upon the wall of the barrier corresponds to an alternating gravitomotive potential $\bar{T}/\Delta x$, which diffracts at the barrier (like affecting a sideways shift of its field lines) in the form of evanescent wave. Along with it, a mechano-capacitive flux Φ_{Cg}/dt develops which is responsible for a flux of displacement momentum through the barrier, an imaginary process that applies as a temporal displacement (tunneling) of matter.

The relation (6-58) could also be re-arranged so as to concern the displacement negative gravitomagnetic field, as follows:

$$(dt/\Delta x) \ \bar{T} = \Phi_{Cg}$$
$$(dt/L_g) \ \bar{T} = \Phi_{Cg}$$
$$(1/\mu_g) \ \bar{T} = \Phi_{Cg}$$
$$\bar{T}_d = \Phi_{Cg} \qquad\qquad\qquad (6\text{-}59)$$

What this may tell, is that in the case of barrier penetration the imaginary coordinate of the displacement negative gravitomagnetic field applies as a mechano-capacitive shift in the form of flux Φ_{Cg}, driving displacement of negative momentum. While the incident particle's wave function oscillates, its negative energy portion behaves as an evanescent wave, which damps out at short range in the classical world. Consequently, the further away the other end of the barrier lies, the harder to resume the oscillation. This is why the probability of tunneling decreases for wider barriers.

Furthermore, as the displacement effect involves negative energy, the corresponding capacitive phase shift in space should transmit almost instantaneously through the barrier. This should be feasible since the shift rides the loop-like axis XX' of figure 1-1, so it develops loop-like in a symmetric way to as in the electronic case a shift in time develops loop-like across the electric circuit, where all points in the loop engage concurrently (just like all points of a bicycle wheel rotate concurrently, therefore instantaneously with respect to each other). Such instantaneity suggests that the capacitive phase shift should be independent of the length of the barrier region, where that complies with the predictions of the Hartman effect that the tunneling time is independent to the width of the barrier.

The magnitude of the gravitomagnetomotive potential that drives the shift depends on the infinitesimal change of the incident wave (as it becomes incident to the barrier), so the frequency of the matter wave becomes proportional to the extent of the phase shift and consequently to the probability of tunneling as well. The corresponding capacitive displacement phase shift of the matter wave gets to borrow energy from the space domain, and that applies as a backing (opposite force) that constitutes a way of conserving momentum.

The effect of barrier penetration is involved in various phenomena of close proximity, from quantum computing, to nuclear fusion, to biochemical redox reactions like photosynthesis. In all these cases, the displacement of the momentum through the walls of the barrier corresponds to an effect of evanescent wave coupling as per the above concepts.

(b) Reaction causing an enhancing energy supply (powering life)

The gravitomagnetomotive potential may also apply in a complementary way, which allows it to react in an enhancing way. This may be feasible when the shift between the Space and Time metrics is becoming involved in an interaction from the viewpoint of negative energy. The imaginary field coordinate interacts with the imaginary part of particle's wave nature, and these two imaginary entities deliver work in the real domain thanks to the relation $i^2=-1$, and in coordination with a negation the outcome may turn into positive. In this way we get energy flow from the imaginary domain toward the real domain. In such case the corresponding gravitomagnetomotive potential should not carry the prefix "back-", it should carry the prefix "push-" instead, thus we may call it "**push-gravitomagnetomotive potential**" for relevance in the naming. This applies in a similar way to as the magnetomotive potential powers a battery, and may supply energy in certain "exergonic" biochemical reactions, which release more energy in the Real domain than they absorb. Like for instance in **life-powering** interactions that turn food into energy, for which to date there is no adequate theory explaining what drives this sustainable process.

To devise an equation that could describe the process, we may start with a relation which is symmetric to Ampere's law:

$$1/\mu_g \oint T \cdot d\ell_t = p \qquad\qquad\qquad \text{(5-47) or (6-60)}$$

This relation originally describes the gravitomagnetic field around fluid flow (relating to the Bernoulli pressure), so it is not really too close to what we are looking for, however it may be modified as follows:

$$\oint \boldsymbol{T} \cdot d\ell_{\mathrm{t}} = \mu_{\mathrm{g}} \, p \quad \rightarrow$$

$$\oint \boldsymbol{T} \cdot d\ell_{\mathrm{t}} = (L_{\mathrm{g}}/dt) \; p \quad \rightarrow$$

$$\oint \boldsymbol{T} \cdot d\ell_{\mathrm{t}} = L_{\mathrm{g}} \, (dp/dt) \tag{6-61}$$

The above displays a symmetry to the expression of the electromotive force $\oint \boldsymbol{E} \cdot d\ell = -L \, dI/dt$, as well as to its gravitomotive counterpart (6-11) which reads $\oint g \, d\ell_{\mathrm{t}} = -L_{\mathrm{g}} \, dp/dx$, subject to the difference that the right side describes a change of momentum p per *time dt* (instead of per *space dx*). However, this relation seems to be of inductive type, and without any obvious involvement of a (pressure-providing) gravitational-like potential that would make it able to account for powering "exergonic" biochemical reactions. We should therefore try a different approximation.

A relevant point to consider, is that in exergonic reactions the process that takes place causes conventional momentum p and not "displacement momentum" p_{d}. To clarify the difference, the displacement momentum is like the one developing within barrier penetration (as per the term $1/\mu_{\mathrm{g}} \oint \boldsymbol{T} \cdot d\ell_{\mathrm{t}}$), it concerns a shift in the particles' wave nature, generated as the wave function incident to the barrier alternates, where the alternation concerns the displacement gravitational flux $\varepsilon_{\mathrm{g}} \partial \Phi_{\mathrm{g}}/\partial x$. The conventional momentum instead (which is of consideration here) has a different origin. For that one, it holds $p = h/\lambda$ (where that is symmetric to the relation $I = Q/t$ for an electric current) and $h = C_{\mathrm{g}} v$ as presented earlier, so in such way the relation (6-61) becomes much like (6-51) but without being of displacement character, so it reads

$$\oint \boldsymbol{T} \cdot d\ell_{\mathrm{t}} = C_{\mathrm{g}} \, v \, /dt \tag{6-62}$$

In relation to material to be presented later in section 7, elementary particles' quantum nature involves an oscillation of the value of mechano-capacitance $\Delta x/dt$, which resembles a velocity term dx/dt. As indicated earlier, in negative energy the velocity is perceived as a potential, and that seems to refer to the pressure-like potential we are looking for. In such case, the right side of (6-62) seems to reflect a condition of the form $[v \, (dC_{\mathrm{g}}/dx)]$ instead of $[C_{\mathrm{g}} \, (dv/dx)]$, while the units remain the same. This type of oscillation seems to bring an enhancing reaction outcome (instead of an impeding one) and we may call that **self-mechanocapacitance** (for symmetry in the naming to self-mechanoinduction that refers to inertia). And while self-mechano-induction (inertia) associates to a "lag in momentum", the self-mechanocapacitance concerns a "lead in momentum", hence the process applies as in enhancing the flow of the imaginary portion of a particle's matter wave (instead of impeding a change in flow of matter which takes place in the case of inertia). This involves the gravitomagnetomotive potential $(\oint \boldsymbol{T} \cdot d\ell_t)$ in an inverse type of interaction which may engage the imaginary portion of the particle's wave nature (which resembles negative matter). In this case, it seems that the interaction makes use of a particular property of quarks which we shall describe in section 10, and gets quarks interact collectively as if they contained (in part) negative matter. In this case, since the negative matter portion of each quark rides (is living in) negative space, its essence is not imaginary with respect to the version of the gravitomagnetic field which drives the gravitomagnetomotive potential. Hence, the potential $\oint T \cdot d\ell_t$ may apply upon negative matter's real portion (particle-like) of quarks participating in an interaction (not the imaginary, wave-like portion). And since the interaction involves particle nature, the potential should generate *ordinary* momentum (not *displacement* momentum). The field version involved in this case appears to correspond to the following

$$T_d += (1/\varepsilon_g) \oint T = (1/\varepsilon_g)\,(T\cos\theta) \tag{6-63}$$

Note that the angle θ refers to the Weinberg angle, and the cos term refers to the coordinate of energy that is leaking from the imaginary toward the real world. The equation which describes the interaction in this case and engages motion of negative matter is

$$(1/\mu_g)\oint \boldsymbol{T}\cdot d\ell_t = [v\,(dC_g/dx)]$$
$$(dt/{-}\Delta x)\oint \boldsymbol{T}\cdot d\ell_t = [v\,(dC_g/dx)]$$
$$\oint \boldsymbol{T}\cdot d\ell_t = (-\Delta x/dt)\,[v\,(dC_g/dx)]$$
$$\oint \boldsymbol{T}\cdot d\ell_t = -[v\,(dC_g/dt)] \tag{6-64}$$

Mind the difference, that in self-mechanocapacitance (6-64) there is change in mechano-capacitance with respect to time (dt), while in self-mechano-inductance $(\oint \boldsymbol{g}\cdot d\ell_t = -L_g\,dp/dx)$ there is a change in momentum with respect to space (dx). Since the oscillation (actually, the loop-like waviness) of mechano-capacitance deploys with respect to time, if we transfer the term (dC_g/dt) to the left side of the equation it takes the form of reciprocal mechano-permittivity

$$(dt/\Delta x)\oint \boldsymbol{T}\cdot d\ell_t = v \qquad \text{or}$$

$$(1/\varepsilon_g)\oint \boldsymbol{T}\cdot d\ell_t = v \tag{6-65}$$

Actually, it seems that equation (6-65) should be re-written in the inverse direction, since the motional velocity v that arises is not the "cause", it is rather the "effect" of the process. That is,

$$v = (1/\varepsilon_g)\oint \boldsymbol{T} \cdot d\boldsymbol{\ell}_t \qquad\qquad (6\text{-}66)$$

Even though the gravitomagnetomotive potential is a reaction potential, the fact that it interacts with the negative-matter-side of quarks gets it exhibit an enhancing character. And as this interaction may develop collectively by neighboring particles at the atomic scale, it appears as a form of chemical tendency in applying a gravitomagnetomotive potential. Since this potential is radial-like, and the units of $(1/\varepsilon_g)\oint \boldsymbol{T}$ resemble the units of $(1/v)T=g$ as explained previously, the way this potential applies resembles the way of action of the electric field.

The process described through (6-66) where v resembles a potential (pressure), comes as a mechanical analogy to the powering of an electric battery in electronics. In the mechanical case, a corresponding potential is responsible for the process of metabolism, which converts food/fuel to energy. According to existing knowledge, inside most living organisms the foods are broken down and rearranged into molecules which the muscle cells convert into kinetic energy (for instance, in moving or lifting objects). However, while the normal body temperature is not high enough to promote the reactions that sustain life, this is done by substances that operate as catalysts, and **increase the rate of chemical reactions** presumably by increasing the rate and potential at which the atoms, ions, molecules collide and interact. A good example of catalyst concerns **enzymes**, as these facilitate most of the reactions in the body and thus enable proper metabolism.

Metabolic reactions are actually of two basic kinds, catabolic and anabolic. The catabolic reactions associate to the breaking down of compounds of large molecules into smaller ones releasing energy through respiration, for fueling activity. Anabolic reactions concern synthesis of new compounds, by consuming energy. Enzymes allow a reaction to proceed more rapidly, and they achieve an increase in entropy in their environments **by coupling the spontaneous processes of catabolism to non-spontaneous processes of anabolism**. Notice the analogy to the processes of oxidation and reduction taking place in electric batteries, as these are symmetric effects subject to a swapping of parameters of space and time. And just like it works for batteries, a mechanical counterpart process involves the negative energy part of quarks, whose imaginary part interacts and leaks energy toward the real domain by becoming involved in the above enhancing reaction.

The above process takes place in compounds which contain carbon, in most cases hydrogen, and may contain oxygen as well. These are named "organics" as they are dominant in living organisms. Yet it appears that they operate as sort of "batteries" which provide the energy that supports life. Organic compounds involve covalent bonds, where these bonds apply between nonmetals with similar electronegativities, and share electrons between atoms, in their outer orbitals. The shared or bonding pairs have a stable balance of attractive and repulsive forces between the atoms. In such cases, two atoms with equal electronegativity make non-polar covalent bonds, such as H-H.[10] It now seems that the essential forces involved in powering these organic interactions are not electromagnetic-born, they concern the gravitomagnetic interaction.

6.5 Remarks

In relation to what has been considered in this section, the following table may be prepared to display the symmetry between electric and mechanical potentials:

No.	Interaction	Relation	Example
1	Electromotive	$\oint E \cdot d\ell = -L\,dI/dt$	Self-induction (electric coil)
2	Electromotive	$\oint E \cdot d\ell = -\theta\Phi_B/\theta t$	Lenz's law (metal ring)
3	Magnetomotive	$1/\mu_o \oint \boldsymbol{B} \cdot d\ell = V\,dC/dt$	Self-capacitance (electric battery)
4	Magnetomotive	$1/\mu_o \oint \boldsymbol{B} \cdot d\ell = \varepsilon_o \theta\Phi_E/\theta t = I_d$	Displ. current (capacitor)
5	Gravitomotive	$\oint g\,d\ell_t = -L_g\,dp/dx$	Inertia
6	Gravitomotive	$\oint g \cdot d\ell_t = -\theta\Phi_T/\theta x$	Vorticity
7	Gravitomagne-tomotive	$1/\mu_g \oint \boldsymbol{T} \cdot d\ell_t = v\,C_g/dx$	Self-enhancement (biochemical)
8	Gravitomagne-tomotive	$1/\mu_g \oint \boldsymbol{T} \cdot d\ell_t = \varepsilon_g\,\theta\Phi_g/\theta x = p_d$	Displ. momentum (barrier penetr.)

Table 6.1: Electric and mechanical potentials

Similar type of interactions may extend to cover further symmetries:

Analogous interactions along an orthogonal direction describe the propagation of EMWs and GMWs, as described in sections 5.3 and 5.4. In particular, EMWs refer to the transfer of alternations in the metric of negative space (E field related) and negative time (B field related), while the GMWs refer to the transfer of alternations in the curvature of positive time (g field related) and positive space (T field related).

In sections 7 and 8 we shall also figure that oscillations "of" time or space (instead of "with respect to" time or space) concern the essence of fermions. And as that extends to the imaginary domain it may also get to account for spin magnetic moment and spin gravitomagnetic moment.

We shall also exlain why the particular nature of fermions has these particles perceive EMWs and GMWs as if they are composed of energy lumps. These particle-like lumps (as perceived from fermions' reference frame) identify to bosons as the photon γ, the Z^0, the W^+ and W^-, which we shall address in section 9.

Oscillations may also deploy in the form of waviness projected upon space. Like for instance the waviness of the time metric which refers to a particle's gravitational field, and the waviness of imaginary space which refers to a particle's electric field.

Likewise, oscillations (waviness) may deploy on time, referring to the fastness of development and repetition of celestial effects at the cosmic scale.

It turns out that nature is full of symmetries, and so many entities all around us reduce to measurements (units) of energy, space, and time. Like for instance:

Joules / second = units of Power (rate of doing work)

Joules / meter = units of Force (measure of action)

Joules $*$ second = Joules / (1/sec)   = Joules / i sec = Planck constant ref. to
quantization of matter

Joules $*$ meter = Joules / (1/meter) = Joules / i meters = referring to
quantization of charge (considered in section 8)

Energy quantum of the Planck const. / length = momentum   ($h/x = p$)

Charge quantum of Coulomb / time = current   ($q/t = I$)

7

The Electron's Wave Function and its Association to Relativistic Mass

We shall here concentrate on the electron, being a truly elementary particle. The electron's wave function involves a real as well as an imaginary oscillating coordinate:

$$\Psi(x,t)=\cos(kx-\omega t)+i\,\sin(kx-\omega t) \qquad (7\text{-}1)$$

The Cosine part of this function is real, so it should be considered to concern positive energy. The Sine part in turn is imaginary, however this does not restrict this coordinate from playing an almost equally significant role. In fact, we shall particularly consider its involvement in effects of reaction, which concern negative energy.

Let's start from how a particle's wave function arises. For this, we shall refer to the Schrödinger equation. That equation is based on a relation of the general form [Kinetic energy + Potential energy = Total energy] which has as follows:

$$[-(\hbar^2/2m)\ \theta^2\Psi(x,t)/\theta x^2] + [V(x,t)\Psi(x,t)] = [\pm\,i\,\hbar\ \theta\Psi(x,t)/\theta t] \qquad (7\text{-}2)$$

The *first* bracket (on the left) of this equation concerns kinetic energy, and arises from the kinetic energy equation $E = \frac{1}{2}\,mv^2$ by substituting classical momentum $p=mv$ with the momentum as per the de Broglie wavelength $p=h/\lambda$. That is,

$$E_{\text{kinetic}} = \tfrac{1}{2}\,mv^2 = \tfrac{1}{2}\,(1/m)\,m^2v^2 = (1/2m)\,p^2 = \hbar^2/2m\lambda^2 \qquad (7\text{-}3)$$

The *second* bracket of (7-2) concerns potential energy attributed to the influence of external fields on the particle, which dictates how the particle shall be confined and/or accelerated. For reasons of simplicity, we shall here consider dealing with a free particle, like a free electron supposedly located at distant outer space, therefore being extremely far away from any influence of external fields. In that sense we shall not give large significance to the particulars of this potential term at this point. But we shall leave it in place for the moment and come back to it later on.

The bracket at the *right* side of (7-2) concerns total energy, as it refers to the relation of total energy $E=hv$, therefore it is equal to the sum of the kinetic and potential energies. Notice that this side (the right side) of the equation contains an imaginary term i. This should reflect the fact that the particle has a negative form of energy associated with it, which is of equal magnitude to the amounts of positive energy (kinetic + potential). The existence of the corresponding imaginary term does not raise an issue, as we shall see that this may be overcome and **acquire real physical significance if it interacts with another imaginary term**, as the two imaginary terms bring a negative Real outcome, since by definition it holds $i^2 = -1$.

In any case, the involvement of an imaginary term should address a "negative form of energy", and that could get to associate to certain superluminal attributes. In relation to that, we shall seek an intuitive reasoning behind what the imaginary term at the right side of (7-2) may represent, as well as figure out how could a *second* imaginary term engage (since it is needed for interactions' results to turn to real as per $i^2 = -1$). Also, we would need to specify what would that second imaginary term represent. In order to figure out these points, we need to prepare and plug in to (7-2) the analytical expressions of the second space derivative and the first time derivative of the wave function (7-1). These derivatives have as follows (with the wavelength λ having been replaced by the wave number $k=2\pi/\lambda$):

Second space derivative of (7-1):
$$\theta^2\Psi(x,t)/\theta x^2 = -k^2 \cos(kx-\omega t) - k^2 \, i \, \sin(kx-\omega t) \qquad (7\text{-}4)$$

First time derivative of (7-1):
$$\theta\Psi(x,t)/\theta t = \omega \sin(kx-\omega t) - \omega \, i \, \cos(kx-\omega t) \qquad (7\text{-}5)$$

Potential term:
$$V(x,t)\Psi(x,t) = Vo \cos(kx-\omega t) + Vo \, i \, \sin(kx-\omega t) \qquad (7\text{-}6)$$

Substituting (7-4), (7-5), and (7-6) to the Schrödinger equation (7-2), the latter takes the form:

$[-(\hbar^2/2m)\ (-k^2\cos(kx{-}\omega t) - k^2\,i\,\sin(kx{-}\omega t))] + [V(x,t)\ (\cos(kx{-}\omega t){+}i\,\sin(kx{-}\omega t))]$
$= [\pm\,i\,\hbar\ (\omega\sin(kx{-}\omega t) - \omega\,i\,\cos(kx{-}\omega t))]$

$$(7\text{-}7)$$

Here, for that equality to hold for all combinations of independent variables x and t, it is necessary that the coefficients of both the cosine and sine terms be zero.[21] To handle this, we shall put all cosine terms of (7-7) in a separate equation, and do the same for all sine terms as well:

The equation of COSINES is:
$[-(\hbar^2/2m)\ (-k^2\cos(kx{-}\omega t))] + [V_o\cos(kx{-}\omega t)] = [\pm\,i\hbar\ (-\omega i\,\cos(kx{-}\omega t))]$

$$(7\text{-}8)$$

The equation of SINES is:
$[-(\hbar^2/2m)\ (-k^2\,i\,\sin(kx{-}\omega t))] + [V_o\,i\,\sin(kx{-}\omega t)] = [-+\,i\hbar\ \omega\sin(kx{-}\omega t)]$

$$(7\text{-}9)$$

For the coefficients of both equations (7-8) and (7-9) to add up to zero, it requires the following two equations (7-10) and (7-11) to be satisfied at the same time:

$$-(\hbar^2\,k^2/2m) + V_o = \pm\,i\hbar\ (i)\omega \qquad (7\text{-}10)$$

and
$$-(\hbar^2\,k^2 i/2m) + iV_o = -+\,i\hbar\ \omega \qquad (7\text{-}11)$$

Actually, by dividing all the terms of (7-11) by i, it takes the equivalent form

$$-(\hbar^2\,k^2/2m) + V_0 = -+\,i\hbar\ (1/i)\omega \qquad (7\text{-}12)$$

Equations (7-10) and (7-12) are satisfied concurrently only if $i = -1/i$. But this is actually already happening by definition, since for the imaginary term it holds $i^2 = -1$ which **may be re-written as $i = -1/i$, and that is actually what justifies why imaginary entities may project to the real world through their negative reciprocals**. Given the above, equations (7-8) and (7-9) may obtain physical interpretation as follows:

<u>**The equation of COSINES**</u>
The equation of Cosines (7-8) represents properties of the particle which concern the REAL world. As previously, the first bracket on the left side represents kinetic energy, the second bracket on the left side represents potential energy, and the right side of the equation represents total energy. The right side involves TWO imaginary terms, which turns it into Real since it

holds i^2=-1. In such case, we'll seek to interpret what each of the two imaginary terms represents:

i. The "first" imaginary term

This is the one which is next to the **Planck constant h**. It exists because the Schrödinger equation relates a SECOND DERIVATIVE of space to a FIRST DERIVATIVE of time. In this case, the "missing" second derivative (of time) relates to, and accounts for, the 90 degrees phase difference that the Imaginary coordinate axis exhibits with respect to the Real coordinate axis in a graphical representation. This imaginary term i relates to the nature of the Plank constant itself, which appears to have an imaginary origin. In such a case, re-writing the relation (i^2=-1) in the form ($i= -1/i$) suggests that **the imaginary attribute could be represented in the Real world through its negative reciprocal**. This should apply for the parameter of time as well, and seems to represent a physical effect taking place "in time", instead of "with respect to time". And since the right side of (7-8) involves a cycling term, it seems to describe a fluctuation in the metric of time itself. That would concern an effect of negative energy, represented as an oscillation with respect to $-idt$ (instead of with respect to dt). If such an imaginary entity interacts with another imaginary entity, so the outcome may project to the real world, this projection to the real world should appear as a negative reciprocal ($i= -1/i$). Consequently, an oscillation with respect to imaginary time would be represented as taking place with respect to $-1/dt$ (instead of with respect to dt). But then, a division by $-1/dt$ is mathematically equivalent to a multiplication by dt. Thus, the parameter of time goes to the numerator instead of the denominator. Thanks to such a feat, numerous entities exist all around us that are imaginary in origin but get to appear and interact in the classical universe through their negative reciprocals. In fact, a solid example of such a case, is the Planck constant itself, which comprises the fundamental quantum of energy in nature.

In line to the above, h is measured in [Joules · sec], which seems bizarre in terms of units since conventional energy would rather be measured PER time, instead of TIMES time. What seems to hold in the case of the Planck constant, is that it concerns energy measured with respect to imaginary time, and since i dt translates to $-1/dt$ the time gets to go to the numerator. That is, the Plank constant's dimensioning actually corresponds to:

$$[Joules \cdot second] = [Joules / (1/second)] = [Joules / -i \ second] \qquad (7-13)$$

What that tells, is that while in classical physics the power is measured in Joules "with respect to time", **the power associated to the Planck constant should be accounted for as "with respect to imaginary time"** (or, with respect to "negative reciprocal time"). That the Planck constant applies with respect to imaginary time, is critical to the particle's probabilistic nature, as well as its quantization attributes. Since [Joules / (1/second)] is equivalent to the familiar [Joules · second] where time is in the numerator, along with the fact that equations (7-8) and (7-9) have the Planck constant accompanied by a cycling term (kx-ωt), gets to resemble a cycling waviness of the parameter of imaginary time (instead of classical cycling in space). And since the quantum particle concerns an oscillation with respect to $i$dt, this resembles a wave-like oscillation in the metric of time, which compensates for the "missing" second derivative of time at the right side of the Schrödinger equation. Such an oscillation in the time domain is what gives to the particle its **probabilistic nature**, since because of it, the particle gets the unconventional property of living in non-single-valued "progression" of time. That is, the particle's matter appears to corresponds to an oscillation of time (instead of an oscillation of mass "with respect" to time, with angular momentum). This may be loosely visualized as an oscillating phase shift of time, getting a particle (a fermion) to essentially concern an oscillation in non-unitary time, and that is what allows it to interact in a way that seems probabilistic, without disobeying causality.

Notice a point of significance that, as an elementary particle's matter-essence lives in non-single-valued progression of time, we need to figure how this may relate to the gravitational field, as in section 5.1 we considered that the gravitational field concerns the difference in the fastness of passing time between adjacent points in space. This issue was addressed earlier while considering the effect of barrier penetration in section 6.4(a). There, it arose that as per the symmetry between the terms ($\varepsilon_g\ \theta\Phi_g/\theta x$) in mechanics and ($\varepsilon_o\ \theta\Phi_E/\theta t$) in electronics, the oscillating quantity Ψ in a particle's wave function seems to associate to an **"oscillating displacement gravitational flux"**. And as we shall consider ahead, in counterbalance to this oscillation, nature reacts by bringing up an oscillation (actually a "waviness") in time which deploys over space, and that constitutes the particle's static gravitational field.

The particular oscillation in time that comprises the particle's essence, is also responsible for getting the particle to interact with electromagnetic waves as if these waves were made of quantized amounts of energy (photons). The reason is that when an electron interacts with an electromagnetic wave, the non-single-valued (sinusoidal) time metric's essence of the electron's matter, along with a corresponding non-single-valued space metric's essence that we shall later associate to charge, apply as if the electron is perceiving the electromagnetic oscillation taking place over a sinusoidal spread in the time and space metrics concurrently. This gets the electron perceive a classical electromagnetic wave of certain frequency as being a superposition of alternations of higher and of lower frequencies of diminishing strength. (And it similarly holds on how the

EMW perceives the interaction with the particle). Such superposition of alternations of higher and of lower frequencies of diminishing strength is described by **Fourier** transform to yield a single "beat", or lump of energy, and this refers to the "quantum" of energy (associating to the Planck constant). This process is responsible for getting the fields of all bosons behaving particle-like.

The interaction of such a waviness of the electron (which resembles living over a range of time and space metrics) with an electromagnetic wave (whose electric and magnetic fields oscillate in single-valued classical macroscopic time and space as per the value of the corresponding metrics at our cosmic region as depicted in figure 1-1) gets an EMW's fields "engage" in a specific point of the electron's oscillating metrics, and this is what corresponds to what is conventionally called a "**collapse of the wave function**" and "landing" to a definite state (in single-valued time and single-valued space). At this condition the particle's probabilistic nature is reduced to single-valued time and space, and this corresponds to a specific energy state of the particle. So, in this way, classical physics and **the particle nature of the particle refers to "catching" the oscillation of time and space metrics to engage and interact in single-valued positive time metric and single-valued positive space metric**. To the extent that the oscillation of time (and the oscillation of space as we shall discuss later) is considered as a degree of freedom (which is beyond what current formulas consider, as they reflect unitary time and space), we illusively consider such activity to be "random" as per the uncertainty principle (unless special measures are taken to control and confine these oscillations, like via stimulated coherent interactions).

ii. The "second" imaginary term

This imaginary term is the one next to the frequency ω. As this comes next to frequency, it signifies that it relates to the *wave nature* of the particle (not the particle nature). And as in section 5.4 we considered an association of a particle's wave nature to negative energy, such a consideration comes in compliance to the fact that the right side of (7-8) has originated from the IMAGINARY portion of equation (7-1) via the derivation (7-5).

According to the above, that the frequency is imaginary seems to reflect the fact that the particle's wave nature concerns an oscillation of time itself. So instead of an oscillation with respect to time, it rather concerns an oscillation of time itslelf, as per the form of the wave function, this is why an atomic electron of a certain energy state does not conventionally exhibit angular momentum (since this does not involve concentional motion). In this sense, Ψ seems to loosely associate to an entity like an oscillating inductive-like shift, or rather, to an oscillating inductive-like flux. For an atomic electron in particular, this corresponds to a wave-like formation represented as per the familiar lobe-like shapes, which do not need to be moving in space since they include an oscillation in the parameter of time itself to constitute their formation. And that comes in reflection to the fact that the Schrödinger equation relates a *second* derivative of space to a *first* derivative or time.

In counterbalance to the waviness of Ψ in negative energy, another form of waviness should deploy in terms of positive energy. That should form in an orthogonal direction, therefore radially to the particle. Here again, this wouldn't concern a waviness with respect to time, it should rather concern the waviness of time itself. **This actually seems to refer to the particle's gravitational field** which corresponds to a curvature in the time domain, where this curvature resembles the time curve illustrated in figure 1-1 (referring to the gravitational field of a single particle).

An oversimplified practical way to provide a hint of what the corresponding balancing concerns, has as follows:
- A particle's internal essence is accounted for by relating the wave function's **second-space** to **first-time** derivatives, and
- That becomes balanced out by the particle projecting upon the classical world with an inverse relation involving the power of **first-space** (dx) to **second-time** (dt^2) parameters, where that reflects to the units of the gravitational field dx/dt^2.

These two counterbalancing wavinesses, should actually relate to each other through a law that is pretty much symmetric to Faraday's fourth law. In relation to what was mentioned in section 6.5, this balancing should be represented by a formula whose right side would reflect the imaginary-related part of a particle's function $[\pm i\hbar(-\omega i \cos(kx-\omega t))]$ and the left side of the formula would concern a waviness of real time as that deploys over space and identifies to the gravitational field.

This concept complies to the assertion made earlier, that the gravitational field does not correspond to a "bending of spacetime curvature as a consequence to the particle's existence". Instead, it constitutes part of the particle's formation. And while each single particle may have negligible contribution to an overall gravitational field of a celestial body, that negligible field follows the same wave pattern of time with all other constituent particles of the celestial body, with time moving faster the further away from the particle, up to a point at mid-distant universe, then beyond that point the time moving slower until reaching a halt near what we consider classical infinity (point d′ of figure 1-1), turning negative beyond that point. In fact, the reason that a particle's gravitational field lines are radial-like, is because they tend to reach out to the negative pole of the particle toward classical infinity.

Throughout this waviness of the metric of time at celestial scale, it appears that **what we perceive as gravity, is a tendency to get fermions bunch up, so that each particle's own waviness of their time metric, is in-phase with all other particles' waviness of their time metric**. In other words, this could be interpreted in that, fermions tend to bunch up (through gravity) in celestial formations where their particle nature has coherent phase among each other in terms of the metric of time, in some analogy to as bosons tend to bunch up in beams where they are coherent to each other.

The equation of SINES
The equation of Sines (7-9) is copied here for easiness in reference:

$$[-(\hbar^2/2m)\,(-k^2\,i\,\sin(kx-\omega t))] + [V_o\,i\,\sin(kx-\omega t)] = [-+\,i\hbar\;\omega\,\sin(kx-\omega t)]$$

It has similar pattern to the equation of Cosines, but includes an imaginary term in each bracket, and in such case, it should reflect the hidden negative matter portion of electrons. As it represents negative energy, it concerns the part of the particle that engages in interactions through an orthogonal way, like in effects of reaction described in the previous section.

In other words, while in the equation of Cosines (7-8) we figured that the imaginary part of (7-1) may get involved into Real world effects as an "imaginary of the imaginary" interaction thanks to $i^2=-1$, here instead, the equation of Sines (7-9) provides an additional possibility concerning how the real part of (7-1) may get involved in the interactions of reaction character "leaking" energy toward the imaginary domain. It should therefore reflect on how matter's imaginary part engages in effects involving the gravitomotive or the gravitomagnetomotive potentials, like for instance in barrier penetration.

Additionally, the same relation (7-9) should describe the essence of particles that find stable habitat in cosmic regions of negative energy, like neutrinos or antineutrinos, which reach stable form in celestial formations like the pulsars. When in stable form, these particles should attract each other, since they are of same matter sign. For such particles within such regions, the equation of Sines should apply as being an equation of Real coordinates (from the viewpoint of their own habitat), and the equation of Cosines should apply as being an equation of Imaginary coordinates (from the same viewpoint).

A comparison between classical and quantum oscillators
It is of interest to make a comparison between the Real part of the quantum oscillator (7-8) and classical oscillators like that of a mass attached to a spring

(7-14) or that of an electronic circuit oscillator involving a "perfect inductor" (7-15).

Mechanical oscillator: $m\ d^2x/dt^2 = -kx$ (7-14)

Electronic oscillator: $L\ d^2i(t)/dt^2 = -1/C\ i(t)$ (7-15)

One may notice that the classical oscillators are based in a relation of a general form:

[Kinetic Energy] = [Potential Energy] (7-16)

While the quantum oscillator (equation 7-8) is based in a relation of the general form:

[Kinetic Energy] + [Potential Energy] = [Total Energy] (7-17)

While (7-17) is structured differently, there seems to be some inconsistency in that its left side is putting together unequal entities, since the kinetic energy part concerns **matter** while the potential energy part typically concerns **charge** (like an external Coulomb potential). Yet, since the particle is indeed mostly influenced by a Coulomb potential, equation (7-8) works well numerically in getting to describe the deployment of the particle's matter wave.

We shall explore the possibility of reducing the equation of the quantum oscillator to a more raw form, in an effort to approximate the structure of (7-16). We may seek to do that on the basis that the potential energy term on the left side of (7-17) concerns an "external" influence (it describes an external Coulomb potential). So, while it indeed influences the actual shape as well as the acceleration of a quantum particle, it should not constitute a critical factor that is prerequisite for the very existence of a quantum oscillation. For tracing the latter, we may focus on the simplest quantum oscillator case, that of a *free* quantum particle which is sufficiently far away from any substantial influence, so in that case the influence of the potential term tends to null. But if so, would such a condition be adequate enough so as to allow to discard the potential energy term after all?

Notice that in a case where a particle is at rest, or it has a known momentum, then according to the uncertainty principle the uncertainly in the position of its presence could extend over huge (or infinite) distance. In such case, even if an external potential is indeed negligible, by integrating the influence of this potential over the huge extend of space its value would not be zero, it should sum up to a significant value. In fact, in a normal case the value of the total external potential energy should amount for $\frac{1}{2}$ of the total energy. This is so, because based on the following calculation, the kinetic energy of the particle accounts only for one-half of its total energy $E=hv$, since:

$$E_{kinetic}= h^2/2m\lambda^2 = (1/2m)\, p^2 = \tfrac{1}{2}\,(1/m)\, m^2v^2 = \tfrac{1}{2}\, mv^2 =$$

$$= \tfrac{1}{2}\,(mv)\, v = \tfrac{1}{2}\, p\;v\; = \tfrac{1}{2}\,(h/\lambda)\,(\lambda v) = \tfrac{1}{2}\, hv \qquad\qquad (7\text{-}18)$$

Consequently, the *externally*-provided potential energy should respectively account for the other one-half of the total energy hv. Thus, the condition that the particle is very far away from any external influence only tells that the external potential is rather spread over an extremely large span in space. And that spreading could actually be pretty *even* (over space). Still in that case, however, this does not keep the widespread external potential energy from summing up to a value equal to that of the kinetic energy.

In the other hand, for a particle that is free and has no force acting on it (since the influence of any external potential is spread evenly over space), the time-independent Schrödinger equation describing this particle should have a potential term that remains constant: $V(x)$=constant. And if no force is acting on the particle, that assumption should hold regardless of the value of that constant, consequently this assumption should not loose generality even for an arbitrary constant, like $V(x)$=0. As such, we may therefore assume that the *external* potential energy term is not vital for the quantum oscillator to be able to oscillate at all, in the first place. That is, even though it is essential for defining the characteristics of the oscillation, it does not seem to be imperative for a quantum oscillator to be able to exist. Based on that, we may proceed to waive the Potential Energy term from the left side of (7-17), based on the assumption that this term is not vital for the simplest form of a "raw" quantum oscillation to be able to oscillate.

Then of course, we need to adjust accordingly the right side of (7-17). Since we waived ½ of the total energy from this equation's left side, we should also do so from the right side too, and reduce that by ½ as well. What we would then get, is the equation (7-19), which is a reduced version of (7-8) since it does not contain the "external Potential Energy" term at the left side, and the energy at the right side is accordingly reduced by one-half. Then, by plugging-in to (7-19) the values of the second derivative of space and the first derivative of time, we get (7-20):

Quantum oscillator (reduced version):

$$[-(\hbar^2/2m)\; \theta^2\Psi(x,t)/\theta x^2] = \tfrac{1}{2}\,[\pm\, i\,\hbar\; \theta\Psi(x,t)/\theta t] \qquad\qquad (7\text{-}19)$$

$$[-(\hbar^2/2m)\,(-k^2\cos(kx-\omega t))] = \tfrac{1}{2}\,[\pm\, i\hbar\,(-\,\omega i\cos(kx-\omega t))] \qquad\qquad (7\text{-}20)$$

The latter two equations (where one could cancel out the ½ factor in both the left and right sides) are now more symmetric to the classical oscillator equations (7-14) and (7-15). One may notice that in both *classical* oscillator cases there is a 90 degrees phase difference between the kinetic energy alternation (at the left side of each equation) and the potential energy alternations (at the right side of each equation). This allows for the average power in such system to be zero, and the oscillation may theoretically keep going on and on (unless there is dissipation of energy). In the case of the *quantum* oscillator instead, we have -again- a 90° phase difference, but here this difference concerns the orthogonality between a real and an imaginary axis, since the kinetic energy term on the left side of (7-19) is of Real origin, while the term at the right side of (7-19) is of Imaginary origin. This may suggest that the right side of (7-19) may describe an "internal potential energy" of negative energy. In such case we do have both a kinetic as well as a potential term for the raw quantum oscillator to be able to oscillate, but **in addition, the energy seems to oscillate between the positive and negative energy** (in the real and the imaginary domains).

The orthogonality between the two sides of (7-19) is in direct compliance to the fact that it equates a second derivative term (at the left side) with a first derivative term (at the right side), and as per general knowledge, when an oscillation is represented by an oscillating curve, the first derivative of it describes the slope (inclination) of the curve, while a second derivative describes the rate of change of that slope. Since the rate of change is maximum when the slope is zero, the first and the second derivatives (of time and space respectively) describe properties which oscillate at a 90 degrees phase difference to each other. This 90-degree difference however, **allows for the average power, along with the total energy, in the system to be zero**, thus the **oscillation may go on and on without need for any external energy**, as long as there is no dissipation of energy. And here, dissipation is avoided by the fact that any change affected on particles is being counterbalanced by another change at the opposite energy, as described throughout section 6. This allows for quantum harmonic oscillators (like the electron) to have really long lifetimes. Furthermore, the relation (7-19) gets to equate a NEGATIVE form of energy (associated to the imaginary domain concerning the right side of the relation) with a POSITIVE form of energy (associating to the real domain concerning the left side of the relation). This in fact, explains how it was ever possible for quantum oscillators (elementary particles) to exist in the cosmos in the first place, **without the need for energy creation, or a need for some external supply of energy supposedly coming from some other hypothetical universe**.

A point of attention is that, for things in the whole cosmos to operate without a need for an external supply of energy, a similar condition like the one that holds for the quantum oscillator might rather need to hold for classical oscillators too. If so, classical oscillators might as well need to have their energy exchanged between positive and negative (imaginary) values. And indeed, while the right side of (7-15) may seem like being real and ordinary, it actually concerns an opposite sign of energy than that of the left side. This is traced in the fact that capacitance needs to interact through its negative reciprocal $-1/C$ (instead of simply C). This negative reciprocation reveals its imaginary nature with respect to the left side of the relation (where inductance L is not reciprocated). Likewise, in the mechanical case described in (7-14) the right side reflects the stiffness of the spring, where stiffness is equivalent to the inverse of flexibility (the negative reciprocal of flexibility), where the flexibility concerns the easiness in affecting a shift in interatomic distances between the atomic particles, an effect involving the gravitomagnetic potential, which is a reaction potential, thus it involves opposite energy. In this sense, the right side of (7-14) also has imaginary origin with respect to the left side of it. Thus, it seems to arise that all of the above oscillators (both quantum and classical) concern oscillation of energy between the positive and negative energy domains.

A relevant point of interest arises through a closer look into what each side of the reduced quantum oscillator equation (7-20) describes:

- The left side of it is what remained after taking away the "external potential energy" term from the left side of (7-8). Therefore, it represents only kinetic energy.
- Regarding the right side, we need to recall that prior to reducing its value by ½, it represented *total energy*. And since that side of the equation was derived from the imaginary part of the wave function (7-1), the corresponding energy should be the sum of two one-halves of energy, one-half of "imaginary kinetic" and another one-half of "imaginary potential" energy. Now that we have reduced the energy of the side of the equation by ½, it must be representing only the one, out of these two one-halves of energy (instead of total energy). It turns out that it should represent the one-half concerning the "imaginary *potential*" energy, since potential energy is needed for the oscillator to be able to oscillate in the first place (independently to the wave form's shape or its overall acceleration which could depend on other, external potentials). The potential energy represented at the right side of (7-20) should be considered "internal" to the particle's quantum oscillator "system", and also, due to the co-existence of a second imaginary term i at that side of the equation, that should have an actual footprint to the real world.

The physical footprint of the particle's internal potential energy may be figured by considering that the right side of (7-20) arose from the imaginary part of (7-1) via the calculation of its first time derivative (7-5). Due to this derivation, the potential energy should associate to an effect of time. And this could have a dual reading:

- While it misses to relate to a second derivative of space, it carries an imaginary term, so it seems to concern an effect of imaginary time. This is traced in the fact that it engages with the Planck constant, which is measured in units of Joules/(1/sec) which refers to Joules/idt. Here the imaginary term accounts for a negative reciprocation of the parameter of time, on the basis that $i^2 = -1$ so $i = -1/i$.

- As for the portion $(- \omega i \cos(kx - \omega t))$ at the *right side* of (7-20), this seems to refer to a waviness in the metric of time, in this case "positive time" due to the co-involvement of two imaginary terms in that side of the equation. That seems to address a waviness of positive time, deployed over space, therefore having the form of a potential. That potential is no one else but the particle's **gravitational** field. And while it appears as a field that is external to the particle, it rather seems that a particle's own gravitational field comprises an essential part of a particle's quantum oscillator, as it serves as the potential that allows for the particle's raw quantum oscillation to take place in the first place.

A relevant point of attention is that, since above we reduced the energy at the right side of (7-20) by one-half, and figured that what we are left with concerns "potential energy in imaginary time", one may wonder what would the other one-half correspond to (the one which we waived), and where would its footprint be. While the original total energy term $E = hv$ is too general, this question may be open to more than one interpretation.

➢ In one interpretation, we may assume that this "other one-half" refers to "kinetic energy in imaginary time, referring to the oscillation in the metric of time which accounts for the particle's matter and the associated quantization of energy as per the aforementioned Fourier transformation process.

➢ In an alternative interpretation, we may assume that this "other one-half of energy" may associate to the particle's charge, which we shall consider in the next section. This option matches much better with the fact that in the above approach to explaining (7-20) we have taken out the "potential energy" portion from the left side of the equation which typically relates to the Coulomb potential. So, in this way we left only the "kinetic energy" term which accounts for one-half of the particle's total energy of $E = mc^2$. The other one-half could alternatively be allocated so as to concern charge (to be considered later).

Conclusively, the energy of a raw quantum oscillator of a fermionic particle seems to alternate **between kinetic energy over the space domain, and opposite potential energy over the time domain**. This allows an oscillation whose energy sum for an elementary particle (e.g. an electron) is zero, meaning that it "costs nothing" to nature for that particle to be in existence.

Relativistic mass

A point that needs to be resolved is how the **relativistic increase in mass** takes place when a particle moves at a relativistic velocity. Here, the wording "increase in mass" is misleading as the effect does not have to do with the particle exhibiting more matter, or becoming larger in some way. Regarding the **inertial** behavior of matter, in section 6.2 it was explained that inertia arises due to the (inductive-like) phase shift between the curvatures of Time and Space depicted in figure 1-1. This (non-local) phase shift extends widely in the universe, and is here postulated to refer to the Higgs field that permeates spacetime. The exact value of this shift (when measured in degrees of angle) appears to match to the Weinberg angle $\theta_w=30°$. This value is here specified indirectly through re-interpretation to older quantitative predictions, according to which, a particular surface exists outside a black hole's event horizon, at a distance of 1.5 times the radius of the black hole [1]. At either side of this surface the centrifugal force points in opposite direction, and light bends in the opposite direction. The new model now re-interprets this surface as a Space node, corresponding to point b of figure 1-1. That this nodal surface is located at 1.5 the radius of the black hole (or, 50% of the radius away) may be interpreted as a phase shift of 30° (the 1/3 of this 1.5, or, the 1/3 of the orthogonal 90°) which exactly corresponds to the Weinberg angle. As shown in figure 7.1, the existence of this shift gets the energy spent to push and accelerate an object (diagonal arrow) to be split in two coordinates: The "real" (horizontal) coordinate does work in accelerating the object, while the "imaginary" (horizontal) one does not do work and only reduces the efficiency of the action.

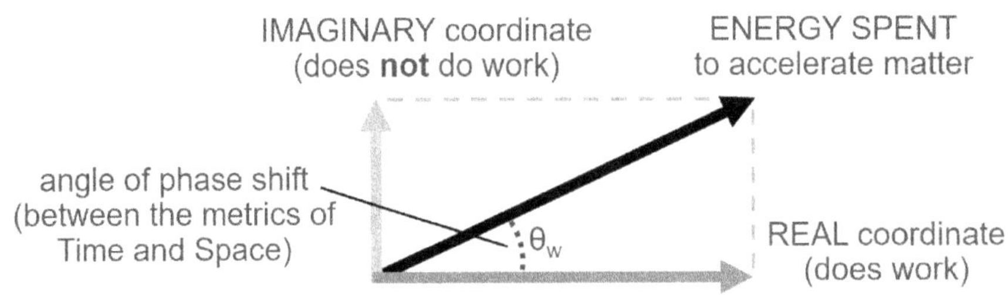

Figure 7-1: How inertial impedance functions

190

In this way, this phase shift gets to "dress" matter with the property of inertia, by affecting impedance (reaction) to a change in the motional status of matter, in accordance to the relation (6-12) which reads $-(1/\mu_g)\oint g=dp/dx$, where the term μ_g involves the inductive-like phase shift L_g. Notice its symmetry with the electric case where an inductive shift between current and voltage oscillations causes inductive impedance which reduces the efficiency in the operation of electric engines. In both cases, however, the relations are not-relativistic.

Since a phase shift is a wave-related effect (applies between the wavinesses of Time and Space), it associates to the particle's *wave nature*, and particularly its wavelength, which associates to its **momentum** as per the wave-nature relation $p=h/\lambda$, and *not* to its mass alone. (Actually it similarly holds with respect to the uncertainty principle, whose relation $\Delta x \Delta p >= h/2\pi$ concerns *momentum*, not mass). What this tells, is that the inertial impedance reacts a *change in momentum* p (and wavelength), not a change in velocity alone. However, when we consider a particle moving at a relativistic velocity, we are accustom to refer to the particle in the conventional sense concerning its *particle nature*, where momentum equals "mass times velocity". Since velocity is considered a property that is measured externally (meaning that it does not concern an intrinsic property of a particle), this makes us think -rather illusively- that the relativistic effect concerns the mass alone. But that is not correct as the relativistic effect affects the particle's wave number, therefore its momentum h/λ, and along with that comes an associated relativistic increase in inertial impedance. When a particle moves at relativistic velocity, the Lorentz term brings the consequence that more conventional power is needed for bringing the same acceleration result (something analogous to an electric engine operating at a bad coefficient of performance, in which case more energy is required for doing the desired amount of work). In effect, when the velocity gets relativistic, space contraction of the wavelength brings an equivalent of a larger "load" and corresponding worsening "coefficient of performance". So, an external potential becomes less effective in accelerating the particle due to the Lorentz term, thus more power is needed to change the motional status (the momentum) of the particle. And this corresponds to what is we perceive as a relativistic increase in the particle's mass.

The same notion similarly applies for matter's **gravitational** behavior. As described in section 5.1, gravitational acceleration develops due to a Doppler-like effect in time which relaxes and releases a gravitational potential. We used the example with the accelerating rocket-ship, where time "moves" slightly faster in the front part than in the rear part by a factor of $(1 + gd/c^2)$, where g is the acceleration and d is the distance between the two clocks [12]. But that factor is only good for a non-relativistic velocity, where the term v^2/c^2 tends to zero.

In a relativistic case, the difference between the clocks at the front and the rear of the accelerating object is subject to the Lorentz term $1/(1-v^2/c^2)^{1/2}$, which yields larger difference. This larger difference corresponds to a larger force, and that provides the illusion of a relativistic increase in mass.

A question of interest is if a similar effect of relativity applies for a particle's charge as well. According to existing theory, when charge is accelerated, it is NOT subject to a relativistic increase. Thus, the charge of a particle that moves at relativistic velocity remains unchanged. However, in this case we need to realize that "charge" is the electric counterpart of "matter" (not mass, neither momentum). So, it should indeed be expected that charge is not subject to relativistic increase, just like matter alone is not subject to a relativistic increase (since h is steady). What changes when a charged object moves at relativistic velocity, is that the object's length becomes space-contracted (gets shorten) by a factor of $1/(1-v^2/c^2)^{1/2}$. In such case, the same amount of charge fits over a shorter length, so what has changed, is the **charge density** of the distribution of charges [14]. In particular, the distribution of charges changes in similar way to the relativistic change in momentum of a particle. In accordance to the present theory, the electric counterpart of momentum, is the electric current which concerns the amount of charge carried with respect to time $I=Q/t$. If we consider that the relativistically moving object (supposedly a wire containing charges) has a larger density of charges along the direction of motion due to its relativistic space contraction, this resembles a larger effective current carried per time. Or, looking into the same effect from a complementary perspective which refers to time, we could consider that the relativistic time dilation (effective decrease) in the denominator of Q/t yields an effective increase in I. In either case, this concerns a relativistic increase in current, in symmetry and analogy to the relativistic increase in momentum in mechanics (as per the particle's wave nature).

Here again, the relativistic increase applies only toward the direction of motion, since the velocity in the transverse direction is zero and hence there is no relativistic contraction. Even though the relativistic effect may be approximated differently depending on the frame of reference, the end result is same. And by analogy to mechanics, if we were to affect a change in this relativistic current, that would be subject to an increased self-induction, in a similar sense to as a change in velocity of relativistically moving matter is subject to a relativistically increased inertial impedance. What that tells, is that contrary to conventional notion, a current is subject to effects of relativity in a similar sense to as momentum is.

Quantization of intrinsic Spin, mechanical-wise

Spin-like motion/flow develops all around us in nature. From vortex flow, to the motion of celestial objects, to inductive currents in circuit loops, and more. In the case of a particle like an electron, spin does not only concern electric & magnetic properties, it also concerns the particle's *mechanical* properties. To approximate the process quantitatively, one could devise a Schrödinger-like relation, by following similar principles to the ones adopted through (7-19). Such an equation should have a kinetic energy term at the left side (in positive energy), and a potential energy term at the right side (in negative energy). To devise a kinetic energy term, we may initially consider that conventional rotational energy is described by the relation $W=\frac{1}{2}\,I\omega^2$, where I is the moment of inertia around the axis of rotation (not to confuse with the electric current I), and ω is the angular frequency. The quantity $I\omega$ corresponds to the angular momentum L as per the relation $L=I\omega$ (not to confuse L with inductance L), in that sense the kinetic energy term could look like $[-\frac{1}{2}\,I\,\omega^2] \rightarrow [-\frac{1}{2}\,(1/I)\,(I^2\omega^2)]$ $\rightarrow [-\frac{1}{2}\,(1/I)L^2] \rightarrow -(L^2/2I)$. That could lead to a Schrödinger-like relation of a conceptional form as per (7-21). Here, it holds $L=m\upsilon r= pr =(h/\lambda)\,r=hr/\lambda$.

$$[-(L^2/2I)\,\theta^2\Psi(x,t)/\theta x^2]=\tfrac{1}{2}[\pm\,i\,L\,\theta\Psi(x,t)/\theta t] \qquad (7\text{-}21)$$

Note however, that in the classical physics context the angular momentum of a system tends to remain constant, so if the moment of inertia changes (e.g. decreases) the angular velocity must also change in the other direction (e.g. increase) for the product of the two to remain steady; this is exactly what holds in the example of a spinning skater that pulls his/her arms and gets to rotate faster. But unlike classical spinning, the *intrinsic* mechanical spin of an electron concerns a different geometric setting, as it involves the imaginary portion of the particle. To approximate that, we shall go through a parallel to electronics: The flow of current in a metal ring comes along with a magnetic flux Φ_B through the ring, and we shall assume a similar condition for a classical gravitomagnetic flux Φ_T through the whirl's center-hole. We may likewise devise a kinetic energy term that uses the gravitomagnetic flux Φ_T instead of angular momentum L, and uses the mechano-inductance L_g instead of the moment of inertia. In a classical-mechanics case, the circular-like flow of matter gets the gravito-magnetic field "lines" being denser inside the whirl and less dense outside of it. This change in density creates a (centripetal) gravito-magnetic potential, and that is reacted through the property of mechano-inductance L_g that causes impedance. In relation to that, the mechano-inductance L_g may take the role of mass. And just like in a metal ring the inductance $L=\Phi_B/I$ is constant and we have an change/alternation in current (and associated magnetic flux around it), in the mechanical analogy we may consider having a constant mechano-inductance $L_g=\Phi_T/p$ as well as a change/

alternation in momentum (and associated gravito-magnetic flux around it). In the *intrinsic* field case of an elementary particle like the electron, in turn, we do not have a classical oscillation of current or momentum. Here, an inverse condition applies, where the magnetic flux Φ_B of the particle is the one that has a constant (quantized) value, while inductance applies as being the one that changes/alternates. And since it should still hold $\Phi_B=LI$, the product LI should stay constant, just like a particle's momentum tends to keep constant in mechanics, or as the quantity CV stays constant and the particle's charge $Q=CV$ has the quantized value of 1e. Likewise, for the mechanical analogy it is the gravitomagnetic flux Φ_T of the particle that has a constant (quantized) value, while mechano-inductance applies as being effectively alternating. And since it should still hold $\Phi_T=L_g p$, the product $L_g p$ should stay constant (just like a particle's momentum tends to keep constant in mechanics). In that sense we may prepare a "kinetic energy" term for a Schrödinger-like relation, having the form of (7-22). That in turn, may find its place in a reduced Schrödinger-like relation of a form (7-23). Notice that in this case (of the mechanical *intrinsic* spin), the gravitomagnetic flux Φ_T resembles the angular momentum L of the classical (*orbit*-like) case.

$$[-\tfrac{1}{2}\,L_g\,p^2] \rightarrow [-\tfrac{1}{2}\,(1/L_g)\,(L_g^2 p^2)] \rightarrow [-\tfrac{1}{2}\,(1/L_g)\,\Phi_T^2] \rightarrow -(\Phi_T^2/2L_g) \qquad (7\text{-}22)$$

$$[-(\Phi_T^2/2L_g)\,\partial^2\Psi(x,t)/\partial x^2] = \tfrac{1}{2}\,[\pm i\Phi_T\,\partial\Psi(x,t)/\partial t] \qquad (7\text{-}23)$$

The right side of (7-23) appears to concern an intrinsic gravitomagnetic field variant, having radial-like field "lines", looping through the very distant universe, where beyond a certain point it should become imaginary and hidden from direct perception. This potential seems to be the one that engages electrons in Van der Waals forces, which concern a mechanical spin-polarization interaction of nearby particles, and refers to the action of what is known as exchange forces. These behave much like tiny bar magnets of "negative-gravitomagnetic" field. Due to the inversion of parameters between the gravitomagnetic and electromagnetic interactions, the corresponding forces point in opposite direction (than the antimagnetic ones), attracting particles of same spin negative gravitomagnetic moment of ½, and repelling away ones of opposite spin. (Recall that such an opposite direction of force also holds in that two masses attract each other, while two charges of same sign repel each other). In this sense, the intrinsic gravitomagnetic potential should apply in opposite direction to the one concerning the Pauli exclusion principle (where the latter potential is repulsive instead of attractive, and is attributed to the intrinsic magnetic field as explained earlier). As the corresponding potential has imaginary character, it has short range, since beyond that range it becomes evanescent, hence these forces do not extend further between the molecules. The imaginary nature of the intrinsic gravito-magnetic field gets spin to project to the real world through a sine coordinate, where the angle θ identifies to the Weinberg angle which represents the 30° phase shift between the Space and Time curves of figure 1-1. In this case, the projection toward the real world is subject to a term Sin(30°) = ½, which lets it behave as if the particle has **gravitomagnetic moment of** ½, providing to it the characteristics of a fermion. The remaining ½ value seems to refer to negative energy, and superluminal characteristics.

The road to new equations (conceptual)

The above concepts may lead the way towards the formulation of new equations describing elementary particles. For this, we would need to bridge the essential difference between quantum-particle-related oscillations and classical-wave-related oscillations, where that may be approximated as follows:

In Maxwell equations we typically assume steady L (hidden within the permeability constant) and steady C (hidden within the permittivity constant) and alternating E and B fields (or fluxes). Likewise, equations of gravitomagnetism assume steady L_g (hidden within the mechano-permeability constant) and steady C_g (hidden within the mechano-permittivity constant) and alternating g and T fields (or fluxes). In the quantum case, instead, matter waves concern exactly the opposite, steady g and T fields and alternating L_g and C_g. These conditions get classical waves to be perceived by particles as quantized lumps as considered earlier, which constitutes the underlying difference between quantum and classical physics. And, as we shall consider in the next section, it similarly applies for the case of charge.

Consequently, the parameters L_g and C_g may apply as oscillating fields (or fluxes), which should also come in displacement versions L_{gd} and C_{gd}, as per the following conceptual relations (7-24) and (7-25). Each of these fields may have two components, a Real one and an Imaginary one, which reflect the particle's positive energy and negative energy parts. Of these components, one should deploy loop-like and the other radial-like, similarly to as it holds for their electromagnetic counterparts.

$$C_{gd} = \varepsilon_h \, (C_g \cos\theta + i \, C_g \sin\theta) \qquad\qquad (7\text{-}24)$$

$$L_{gd} = 1/\mu_h \, (L_g \cos\theta - i \, L_g \sin\theta) \qquad\qquad (7\text{-}25)$$

Notice that the above relations should involve a new type of permittivity-like and permeability-like terms. Unlike the electric case where ε and μ involve a shift in time $\pm\Delta t/dx$ associating to a fixed capacitance or inductance, in the case of matter the terms ε_h and μ_h (where the subscript "h" indicates a version concerning matter) should rather involve **a displacement in the particle's gravitational and gravitomagnetic fields** (to be considered in section 10). These terms could have as -conceptually- suggested in table 7.1 third column.

Gravito-gravitomagnetic	Electro-Magnetic	Mattero-Spin	Chargo-Spin
$\varepsilon_g = C_g/dt$ $= +\Delta x/dt$	$\varepsilon = C/dx$ $= +\Delta t/dx$	$\varepsilon_h = g/(1/dt)$ $= dx/dt$	$\varepsilon_Q = E/(1/dx)$ $= dt/dx$
$\mu_g = L_g/dt$ $= -\Delta x/dt$	$\mu = L/dx$ $= -\Delta t/dx$	$\mu_h = T(1/v)/(1/dt)$ $= -dx/dt$	$\mu_Q = Bv/(1/dx)$ $= -dt/dx$

Table 7.1: Permittivity-like and Permeability-like regulators

The terms shown in table 7.1 should not be considered fixed, they should vary based on positioning of a particle within the waviness of the time metrics shown in figure 1-1. Therefore, we could call these terms "regulators". For simplicity, we shall however continue using the name "terms".

The above fields and terms should engage in a set of classical-like relations that are symmetric to Maxwell's fourth law and Lenz law, and describe fermions. One such equation could have its one side describe a waviness in time which concerns the essence of matter, while the other side should describe the waviness in the metric of time as that deploys in space which identifies to a particle's gravitational field. In this case, the values of the permittivity-like and permeability-like terms would define the type of fermion, as in relation to material to follow in section 10 **the phase shift involved in these terms may provide for the difference between a lepton (e.g. an electron or positron or neutrino) and a quark**. Likewise, a similar relation should describe a mechano-capacitive waviness and a counterbalancing gravitomagnetic field at the other side.

In relation to further material to be considered in section 9, similar relations should more appropriately apply at cross-interaction level. For instance, instead of involving an alternation of C_g and L_g, they should describe the alternation of fields from both the gravitomagnetic and electromagnetic interactions, for instance L_g and C, where that seems to reflect the need for coexistence of matter and charge in forming an elementary particle. That in fact, makes it reasonable to consider the co-existence of matter and charge in what we may call a **"chargo-matter"** (or "electro-gravitational") interaction, of which the "mediators" are simply, the **fermions**.

8

Charge, and its Association to the Particle's Wave Nature

Like previously, we shall here focus on the electron, a truly elementary (not composite) particle. Charge has certain similarities with matter, for example, it comes along with a radial-like field (the electric field) just like matter comes along with the radial-like field (the gravitational field). Or, a change in the motional status of charge becomes subject to electromotive impedance, in certain analogy to as matter is subject to inertial impedance. On the other hand, while matter exhibits wave-like properties, charge is considered to follow the wave form of matter, and not to exhibit independent wave-like properties of its own, since the Schrödinger equation does not seem to predict a "charge wave" in some way. A corresponding property of waviness for charge may however be seek through a symmetric process to the one used for obtaining a particle's matter wave.

Above we have figured that effects of gravity are symmetric to effects of electricity via the swapping of the parameters of time and space. It is therefore reasonable to expect that a similar symmetry should hold between the essence of matter and the essence of charge, relating to each other through a similar swapping of parameters. In this case, however, the symmetry appears to break due to the fact that matter arises by relating a first derivative of *time* to a second derivative of *space*. In relation to this, matter gets to have orthogonal properties with respect to charge, where for charge we may perceive the positive (+) pole and the negative (-) pole and may not perceive a positive anti-pole (anti+) and a negative anti-pole (anti-), while for matter we may perceive positive matter (+) and positive antimatter (anti+) poles and may not perceive negative matter

(-) and negative antimatter (anti-) poles, as was depicted in figure 4-1. Due to such orthogonality we may not perceive charge waves in the same way we perceive matter waves, actually their behaviors deploy in opposite domains (for one of them in the space domain and for the other in the time domain).

Such orthogonality however allows for approximating charge through a process which uses the derivatives of space and time in a complementary way to the one used in obtaining the Schrödinger equation in the case of matter. For the cosmic region $d' - b$ of figure 1-1 (which corresponds to our familiar classical universe), this complementary process would relate a FIRST derivative of SPACE, to a SECOND derivative of TIME (instead of a first derivative of time to a second derivative of space which was the case for matter). Such a relation should therefore look much like a Schrödinger equation having the space and time parameters swapped.

The conventional Schrödinger equation is based on a general relation of the form [Kinetic energy + Potential energy = Total energy], so this now should be re-adjusted to reflect on charge. While effects of charge and matter seem to be subject to a swapping of parameters of space and time, it requires cautiousness in figuring out how does this concern the Kinetic and Potential energies involved. We shall start by tracing what a "Kinetic energy in relation to charge" may correspond to. Staring with a classical-like notion that would be symmetric to the relation $E_{kinetic}= \frac{1}{2} mv^2$ of mechanics, we shall explore what could be playing the role of **mass**, and how **velocity** would become involved given the inversion of parameters:

- We may begin by pointing out that **mass** corresponds to "matter dressed with the property of inertia". Inertia, in turn, corresponds to an impedance that gets "energy being consumed without doing work" in accelerating a particle. And the value of such impedance depends of the extend of the shift between the space and time curves of figure 1-1, since the larger the shift, the less the efficiency in accelerating matter. A corresponding analogy in electronics may be traced in the property of **capacitance**, in the sense that a capacitive phase lead of a voltage oscillation with respect to a current oscillation gets part of the energy applied to be consumed without generating work, thus affecting an impedance to a change in current. In that sense, we have a first indication that capacitance (which in the conventional sense relates to how much charge can be contained in a capacitor) may play a somewhat similar role in electronics with the role that mass does in mechanics.

- As far as **velocity** is concerned, in the context of electronics we might rather be looking for a property that is physically different to the conventional velocity we are classically accustomed to. Considering that electronics differ from mechanics through a swapping of the parameters of space and time, we could rather expect a charge-related analogy to velocity having units much like dt/dx (instead of the conventional dx/dt). Or, looking into the same point via a complementary approach, we may consider that, since the electric interaction is orthogonal to the gravitational, this implies that either of these two interactions is imaginary with respect to the other. Hence, if properties related to electronics are imaginary with respect to properties related to mechanics, that should rather suggest that from the point of view of electronics the parameters of both space and time should project to the real world through their negative reciprocals (as per $i = -1/i$). Thus, a velocity-related term would rather correspond to $(-1/dx)/(-1/dt)$ instead of the conventional dx/dt, and that leads to the same notion, of the velocity in electronics effectively corresponding to a property whose units refer to dt/dx.

Tracing on what the physical essence of dt/dx could correspond to, and how it would be perceived by us (where "us" refers to our matter-related classical physics point of view), we may refer to the previously presented units of the electric field being $dt/d(-x)^2$. The electric field relates to voltage (e.g. the voltage between a capacitor's plates) as per the relation $\Delta V=Ed$ where d is the separation between the capacitor's plates. According to that, the units of (oscillating) voltage would correspond to $[dt/d(-x)^2][dx] = dt/dx$. We shall therefore take this note as a base, to suppose that in the context of electronics, voltage dt/dx assumes a role similar to the one that velocity dx/dt does in mechanics.

On a point of interest, as previously explained, the reciprocal of velocity dt/dx is perceived in the classical world to correspond to voltage, the same should apply in reverse as well, where **what we perceive as conventional motion dx/dt should be perceived as a potential from charge's point of view**. This is actually why the classical motion (dx/dt) of charge forms a potential around it, which was explained to identify to the magnetic potential around a current.

According to the above, the "Kinetic energy in relation to charge" could be approximated by a relation of the form $W=\frac{1}{2}CV^2$ (where C has a role like that of the inertial impedance of mass in mechanics, and V has a role like that of velocity). This is an existing relation, however, the term $\frac{1}{2}CV^2$ does not seem to have appropriate form for use in a complementary-to-Schrödinger equation, as it rather associates to a "particle-nature" of charge (very much like the term $\frac{1}{2}mv^2$ does in mechanics), while we need an expression that would refer to the "wave nature" of charge (in analogy to the kinetic energy term being $\hbar^2/2m$ in mechanics). Furthermore, while the term $\frac{1}{2}CV^2$ is appropriate to describe alternations of voltage in a circuit involving a capacitor, where it holds $C=Q/V$ and capacitance C is constant, in the case of an elementary particle like the electron it is the charge Q that has a constant (quantized) value, not the capacitance. And since it should still hold $Q=CV$, the product CV should stay constant, in some analogy to as a particle's momentum $p=mv$ tends to keep constant in mechanics. (Actually, for such a comparison it is not exactly p, it is rather $h=p/\lambda=(m/\lambda)v$ that remains constant and quantized, where h is the Planck constant, and (m/λ) can be considered to reflect how much inertial mass is contained per wavelength λ, in a similar sense to as capacitance reflects how much charge can be contained in a capacitor).

In that sense, it is more appropriate to look into any comparisons based on the wave nature of particles. And while the waviness of matter is counterbalanced by a radial potential that concerns the gravitational field, the waviness of charge is counterbalanced by an orthogonal formation of V whose field lines identify to the electric field, which extends radial-like and loops through some point toward very distant universe where the field "lines" turn imaginary so their "return" portion is hidden, and we directly perceive only the real portion of the field. Furthermore, while the electric field was explained to correspond to the curvature of "negative-reciprocal space", that holds only from the reference point of mechanics. While from the reference point of electronics, the electric field corresponds to the curvature of "plain" real space. This is due to the aforementioned orthogonality between mechanics (the gravitational interaction) and electronics (the electromagnetic interaction). So, the electric field's existence appears to come in compliance to a Schrödinger-like equation relating a first derivative of space to a second derivative of time.

Based on the above, we shall go through the following modification in obtaining a kinetic energy coefficient in relation to charge, which is "quantum-physics-compatible" and appropriate to use in describing the particle's "kinetic energy in relation to charge":

$$[-\tfrac{1}{2}\,C(\Delta V)^2] \rightarrow [-\tfrac{1}{2}\,(1/C)(C^2 V^2)] \rightarrow [-\tfrac{1}{2}\,(1/C)Q^2] \rightarrow -(Q^2/2C) \qquad (8\text{-}1)$$

Notice how symmetric the term $(Q^2/2C)$ is to the kinetic energy coefficient $(\hbar^2/2m)$ of quantum mechanics. Actually, here we have a replacement of the quantized value of h, by another quantized value, the one concerning the electron's charge Q. With this on hand, we may now come back into formulating an equation of the general form [Kinetic energy + Potential energy = Total energy] for the case that concerns the electron charge, as follows:

$$[-(Q^2/2C)\ \theta^2\Psi(x,t)/\theta t^2] + [V(x,t)\ \Psi(x,t)] = [\pm\,i\ Q\ \theta\Psi(x,t)/\theta x] \qquad (8\text{-}2)$$

This may be worked out in a similar way to as is done in the actual Schrödinger equation, that is, to calculate the second time derivative of Ψ as well as the first space derivative of Ψ, of equation (6-1). Actually, **the expression of Ψ could come in variations, these variations involving the swapping of the cosine and sine terms and the parameters of space and time in order to properly account for each different lepton (an electron or a positron or a neutrino or an antineutrino) in each region of spacetime** depicted in figure 1-1.

And it similarly holds for the potential term as well (second bracket on the left side of 8-2), depending on the acting fields, however in the present case this term may be kept as is, considering that the potential energy typically concerns the Coulomb potential in any way. In such case the derivatives have as follows (with the wavelength λ having been replaced by the wave number $k=2\pi/\lambda$):

First space derivative:
$$\theta\Psi(x,t)/\theta x = k \sin(kx\text{-}\omega t) + k \, i \cos(kx\text{-}\omega t) \tag{8-3}$$

Second time derivative:
$$\theta^2\Psi(x,t)/\theta t^2 = -\,\omega^2 \cos(kx\text{-}\omega t) - \omega^2 \, i \sin(kx\text{-}\omega t) \tag{8-4}$$

Potential term:
$$V(x,t)\Psi(x,t) = V_o \cos(kx\text{-}\omega t) + V_o \, i \sin(kx\text{-}\omega t) \tag{8-5}$$

By plugging in the corresponding second time derivative and first space derivative of Ψ to (8-2), we may then separate the Cosine and Sine terms into two separate equations in a similar manner to as was done with (7-8) and (7-9). These may indicatively have as follows (depending on the variations of Ψ and V_o used in the calculations):

Equation of COSINE terms:
$$[-(Q^2/2C)\,(-\omega^2 \cos(kx\text{-}\omega t))] + [V_o \cos(kx\text{-}\omega t)] = [\pm\, iQ\,(-\,k\,i\cos(kx\text{-}\omega t))] \tag{8-6}$$

Equation of SINE terms:
$$[-(Q^2/2C)\,(-\omega^2\,i\sin(kx\text{-}\omega t))] + [V_o\,i\sin(kx\text{-}\omega t)] = [-+iQ\,(k\sin(kx\text{-}\omega t))] \tag{8-7}$$

Similar to what the case was for the matter wave, for the coefficients of both equations (8-6) and (8-7) to add up to zero it requires that $i=-1/i$. But again, this is actually already happening by definition, since for the imaginary term it holds $i^2=-1$.

Using similar reasoning as we did for the case of matter, we could proceed in preparing "reduced equations" for the case of waviness of charge, describing the "raw" quantum oscillator. As previously, this can be done by taking out the V_o term, and reducing the right side of the equation (which represents total energy in relation to a particle's charge) by one-half. What we would then get, has as follows:

Quantum oscillator concerning charge (reduced version):

Equation of COSINE terms:
$$[-(Q^2/2C) (-\omega^2 \cos(kx-\omega t))] = \frac{1}{2} [\pm iQ (- k\, i \cos(kx-\omega t))] \qquad (8\text{-}8)$$

Equation of SINE terms:
$$[-(Q^2/2C) (-\omega^2\, i \sin(kx-\omega t))] = \frac{1}{2} [-+iQ (k \sin(kx-\omega t))] \qquad (8\text{-}9)$$

For both equations (8-8) and (8-9), the left side describes a term of $Q^2/2C$. If we transfer to this term the ω^2 of the cosine or sine terms, then we would get the kinetic energy coefficient to be $(Q^2\omega^2/2C)$. And, similarly to as was done in the Schrödinger equation where the wavelength λ was replaced by the wave number $k = 2\pi/\lambda$, we can replace the frequency ω with its reciprocal **"frequency number"** which we may denote by the letter φ, where $\varphi = 2\pi/\omega$. This concerns the number of cycles (or radians) per unit time (so it is something relevant to "period" instead of an ordinary "frequency"). It seems handy to use this, as in the quantum case we deal with reciprocals of space and/or time. So, when it comes to (8-8) and (8-9) we may substitute ω with $2\pi/\varphi$, and for reasons of convenience we may use $\wp = \varphi/2\pi$, in which case the kinetic energy term now becomes $(Q^2/2C\wp^2)$.

The right side of (8-8) and (8-9) involves the quantity "Qk" which corresponds to the total energy and is symmetric to the "$h\nu$" of the case of the particle's matter. Actually, **the charge Q is quantized for a similar reason to that the Planck constant is quantized** in the case of matter. That reason being that, as the particle involves a waviness of the space metric, it includes a whole range of values of space stretchiness/contraction, so in interactions it applies as a **Fourier** series over an extend of space stretchiness/contraction. **Since in the classical word our perception matches the precise value of the space metric in our cosmic neighborhood, the charge waviness over an extend of reciprocal space "projects" to the classical world as a sum of waves of smaller and bigger wavelengths (of progressive amplitude), and that gets the particle's charge interact with external fields in a quantized fashion.**

The quantity Qk corresponds to the total energy W associated to the particle's charge-related wave nature. In terms of units, the energy is measured in Joules, charge Q is measured in Coulomb, and the wave number k is a measure of units of $1/dx$ since it associates to the reciprocal of wavelength (it describes the number of cycles-or-radians per unit space, so it is something like the "number of wavelengths over a specific distance"). And since it holds $Q=W/k$ in such case the **unit of Coulomb corresponds to [Joules · meters]**. Notice how symmetric this is to the units of the Planck constant [Joules · seconds], the difference being that time has been swapped to space. Just as suggested earlier for the case of the Planck constant, the units of Q would be more appropriate to be written in the form of [Joules / (-1/meter)] instead of [Joules · meter]. In other words, charge can be thought of to concern energy with respect to imaginary space [Joules/(idx)] where the imaginary term accounts for a negative reciprocation on the basis that $i^2=-1$ so $i=-1/i$. As for the negative sign, that comes along with the reciprocation. To the extent that the curvature of space is imaginary to the curvature of time, that may be considered responsible for the electron charge being treated as **"negative" charge** (even though the name of the sign may have been arbitrary in the first place). That may actually be considered to relate to the fact that charges of same charge repel each other while particles of matter of same sign attract each other.

At this point, an interesting reference can be made, to the comparison between classical and quantum oscillators that was presented in the previous section, and more particularly, to the "reduced equation" of the quantum oscillator that was devised in (7-22), where the right part of it was said to account of only ½ of the total energy hv, attributed to internal, imaginary, potential energy. It now seems that the OTHER ½ of energy (which is missing from relation 7-22) could be thought of to associate to the ½ of the energy relating to charge (the one-half of $W-Qk$) at the right side of (8-8).

<u>The equation of COSINES</u>

In symmetry to what was described in the previous section, the equation of Cosines (8-8) represents properties of the particle's charge which project to the real world. The left side represents kinetic energy and the right side of the equation represents a potential energy term, from the point of view of charge. The right side involves TWO imaginary terms which allows it to turn to real as per the relation $i^2=-1$. It is of interest to figure out what each of these two imaginary terms may represent:

i.   The "first" imaginary term, which is next to the quantized value of charge Q, is there because the Schrödinger-like equation concerning charge relates a SECOND DERIVATIVE of time, to a FIRST DERIVATIVE of space. In this case, the imaginary term reflects the "missing" second derivative of space, and is in accord to that an Imaginary coordinate has a 90 degrees phase difference with respect to the Real coordinate.

ii.   The "second" imaginary term is the one next to the wave number k, and it is there because the wave function associated to the particle's charge is a complex entity and the Cosine part that we are dealing with here originates from the imaginary part of the wave function, so it associates to negative energy. And as the corresponding imaginary term stands next to the wave number, that signifies that it relates to the *wave nature* of the particle.

That the wave number is imaginary seems to reflect the fact that the charge's intrinsic wave nature concerns an oscillation of space itself. So instead of an oscillation "with respect to space", it rather concerns space contractedness oscillating as represented by the corresponding wave function, and this is why the electron's charge does not "move" as a classical current in an orbit-like fashion. Furthermore, the reciprocity involved in the wave number gets the oscillation take place as a multiplication instead of division (applies as "times" space instead of "with respect to" space). This multiplicity is what allows for effects of **non-locality** to take place without disobeying causality or the laws of relativity, while the co-involvement of the other imaginary term allows for the projection of interactions' outcome to the Real world. That outcome typically refers to effects of reaction character, in compliance to the negation involved in the relation $i^2=-1$.

Notice that since both positive and negative charge project to the real world (and as a result, the magnetic field associating to the motion of both positive or negative charge is real and applies long range), the capacitive waviness referring to charge is fully real, so the quantized value of the electron's charge equals 1 instead of ½ as it holds for *h*.

And while a formation of waviness may deploy at the atomic scale and reflect an energy state of the electron around the nucleus, conservation of energy demands for a counterbalancing formation to deploy in terms of opposite energy. That should form in an orthogonal direction, therefore radially to the particle. Here again, this wouldn't concern a waviness with respect to space, it should rather concern the waviness of "imaginary space" (negative-reciprocal space). This actually seems to concern the particle's electric field, which corresponds to a curvature in imaginary space (if viewed from the reference point of mechanics).

An oversimplified practical way to provide a hint of what the corresponding balancing concerns, has as follows:

- A charge's internal essence is accounted for by relating the wave function's **second-time** derivative to the **first-negative-space** derivative, and
- That becomes balanced out in the classical world with an inverse relation involving the power of **first-time** (dt) to **second-negative-space** ($d(-x)^2$) parameters, where that reflects to the raw units of the electric field dt/dx^2 as previously considered.

This balancing could be reflected in the form of an interaction formula whose one side concerns the imaginary part of a particle's function oscillating in imaginary space (much like [$\pm iQ (- k\, i\, \cos(kx\text{-}\omega t))$]) so the two imaginary terms bring a real result. In such case, the one side of the formula concerns a waviness of space as that may project over space and identifies to the electric field, and the other concerns the oscillation in space which concerns charge.

In that sense, the electric field should *not* be understood to concern a consequence to the environment due to the particle's existence. Instead, it should be understood to constitute part of the particle's formation (just like it was stated for a particle's gravitational field). And the curvature of imaginary space constituting the electric field (which corresponds to the curvature of plain space if viewed from the reference point of electronics) should follow a waviness of similar type to the ones shown to ride the axis XX' of figure 1-1, stretching negative space more the further away from the particle, but only up to a point, then beyond that point contracting, until reaching some nodal surface at the very distant universe, beyond which it turns sign.

The equation of SINES

The equation of sines (8-9) has similar pattern to the equation of Cosines, but includes an imaginary term in each side of it. In such case, it reflects the hidden anticharge portion of electrons. As it represents negative energy, it engages to the real world through an orthogonal way, hence, in terms of energy it is subject to the Sine coordinate. Through this coordinate it may become involved in transitional effects of reaction (like the electromotive), where it applies at close range, as in the displacement of charge between the plates of a capacitor which was described earlier.

Additionally, the same relation should describe the anticharge essence of particles which find stable habitat in cosmic regions of negative space metric, like neutrinos or antineutrinos.

<u>Quantization of spin and spin magnetic moment:</u>

Electron spin is not a classical physics effect, as in order to produce the measured magnetic moment in classical physics terms, the surface of a charged particle would have to be moving faster than the speed of light. In particular, the electron appears to be twice as effective in producing a magnetic moment as the corresponding classical charged body would do, as it has a spin g-factor of approximately two gs ≈ 2 (with the deviation from the integer value being due to a Zeeman-like influence). It therefore appears that an electron's spin involves the particle's imaginary portion. In fact, this gets the spin g-factor to comply with the electron spin being ½, considering that the projection of imaginary parameters to the real world should come through a negative reciprocal as per $i= -1/i$, and in this case $1/(½)=2$. (Notice that a similar reasoning seems to apply behind the spin 2 of the graviton as we shall consider in the next section).

While spin is not accounted for in the conventional Schrödinger's equation, we may device a way to account for it via similar principles to the ones adopted for the mechanical version of the intrinsic spin in the previous section. For that, we need to estimate -once again- a kinetic energy term for the left side of the corresponding relation (in positive energy), and a potential energy term for the right side of it (in negative energy), for preparing reduced equations of Cosines and Sines. Regarding the kinetic energy term, like in the electric case of above we used capacitance to take the role of mass, and we also used voltage to take the role of mechanical velocity, here we shall seek to devise a relevant kinetic energy term by (initially) parallelizing spin to a loop-like flow of current I through a coil. The flow of current through coil gets the magnetic field "lines" being denser through the hole of the coil, and less dense outside the coil. This change in density creates a magnetic potential, which is reacted through the property of inductance that the coil exhibits and causes effects of impedance. In relation to that, we shall use inductance L to take the role that inertial mass takes in mechanics, and we shall use the current I to resemble the role that velocity takes in mechanics. That may result to a kinetic energy term of the form W=−½ LI^2. But, unlike the case of an electric circuit where the inductance $L= \Phi_B/I$ is constant and we have an alternation in current and in the associated magnetic flux correspondingly, **in the case of an elementary particle like the electron it's the magnetic flux Φ_B of the particle that has a constant (quantized) value,** while inductance should be alternating. And since it should still hold $\Phi_B = LI$, the product $L \cdot I$ should stay constant, just like a particle's momentum tends to keep constant in mechanics, or as the quantity CV stays constant (since the particle's charge $Q = CV$ is quantized as explained earlier). However, while an actual electron does not involve a classical oscillating current, but it involves a "wave-like oscillation" of L, the product $L \cdot I$ should still be kept steady. In such way the magnetic flux Φ_B applies as an analogue of angular momentum, so we may prepare a "kinetic energy" term for a Schrödinger-like relation, as follows:

$$[-\tfrac{1}{2} L\, I^2] \rightarrow [-\tfrac{1}{2}\,(1/L)\,(L^2 I^2)] \rightarrow [-\tfrac{1}{2}\,(1/L)\,\Phi_B^2] \rightarrow -(\Phi_B^2/2L) \qquad (8\text{-}10)$$

That may find its place in a reduced Schrödinger-like relation of a form:

$$[-(\Phi_B^2/2L)\ \theta^2\Psi(x,t)/\theta t^2] = \tfrac{1}{2}\,[\pm i\ \Phi_B\ \theta\Psi(x,t)/\theta x] \qquad (8\text{-}11)$$

Substituting the appropriate derivatives of space and time (depending on the type of elementary particle concerned), the constituent terms may be separated (as previously) into two equations, one concerning the cosine coordinate which represents the Real part of a particle's spin, and a sine coordinate concerning the Imaginary part of the particle's spin. As the spin interaction involves the imaginary portion of the particle's wave function, it should concern the equation of sines, of which both sides are imaginary. The right side seems to relate to the antimagnetic field $\bar{B}$, having radial-like field lines, where that refers to the intrinsic-magnetic field, as described previously in section 5.3 - point 5. The negative energy involved in this field variant lets it apply in the form of a magnetomotive potential, which appears to be the one that engages electrons in forces related to the Pauli exclusion principle, behaving much like tiny bar magnets repelling away particles of same spin antimagnetic moment of ½, and attracting ones of opposite spin.

Even though both sides of a corresponding equation of sines are imaginary, the corresponding attributes of the particle may engage in interactions that bring real results via interaction with another imaginary term associated to an external field version. The sine coordinate's angle identifies to the Weinberg angle which represents the 30° phase shift between the space and time curves of figure 1-1. In this case, the projection toward the real world corresponds to $\mathrm{Sin}(30°) = \tfrac{1}{2}$, which lets it behave as if the particle has **spin magnetic moment of ½**, providing to it the characteristics of a fermion.

The road toward new equations (conceptual)

The above concepts may lead towards the formulation of new equations describing the essence of a particle's charge. These differ from Maxwell equations in that the latter assume steady L (hidden within the permeability constant) and steady C (hidden within the permittivity constant) and changing E and B fields (or fluxes). For a quantum particle like an electron, instead, a charge-related waviness should concern exactly the opposite, steady E and B fields and alternating L and C. Consequently, the parameters L and C should apply as oscillating fields (or fluxes), and should also come in displacement versions L_d and C_d, as per the following conceptual relations (8-12) and (8-13). Each of those has two components, a Real one and an Imaginary one, which reflect the particle's positive energy and negative energy parts. Of these components, one should deploy loop-like and the other radial-like.

$$C_d = \varepsilon_Q \left(C \cos\theta + i\, C \sin\theta \right) \tag{8-12}$$

$$L_d = 1/\mu_Q \left(L \cos\theta - i\, L \sin\theta \right) \tag{8-13}$$

Here, the permittivity-like and the permeability-like terms ε_Q and μ_Q should have values as per table 7.1 right column. In relation to those, new relations could be devised to describe elementary particle's charge, in a symmetric form to Maxwell's fourth law and Lenz law. In one case, an equation's one side should describe a waviness that concerns the essence of charge, while the other side should describe the waviness in the metric of space (or negative-reciprocal space depending on the reference frame) which identifies to a particle's electric field. In this case, the value of permittivity-like and permeability-like terms would define the type of fermion (while, in relation to material to follow in section 10, there is a particular differentiation between a lepton or a quark).

Corresponding relations should also apply at cross-interaction level, involving the alternation of fields from both the gravitomagnetic and electromagnetic interactions, where that would reflect the counterbalancing coexistence of matter and charge in forming an elementary particle. That in fact, makes it reasonable to consider this co-existence in what we may call a "**chargo-matter**" (or "electro-gravitational") interaction, of which the "mediators" are the **fermions**. Such kind of formulation would allow to describe fermions' quantum oscillations in a classical-like form.

9
Concepts on the Electroweak Interaction

The electroweak interaction is considered to be mediated by four particles, namely the photon γ, the Z^0, and the W^+ & W^- bosons, each of which seems to have two field components.

The photon γ

According to theory, the photon is the only massless, and thus the only mediator of the electromagnetic interaction. (This however will be re-considered below, as the photon seems to have an inverse energy cousin which exhibits particular characteristics that seem to make it behave as if it carried mass, the Z^0 boson). The difference between the wave nature and the particle nature of radiation has to do with the reference frame. What we perceive as a classical electromagnetic wave (EMW), concerns the field oscillations and wave propagation with respect to the macroscopic single-valued time metric and single-valued space metric. While, what makes that wave behave as a quantum particle (photon) concerns its interaction with respect to oscillating parameters of time and space which comprise the essence of atomic particles. That is, while in the classical world we perceive an EMW propagating with respect to **unitary**-spacetime (meaning with respect to the specific values of the space and time metrics in our cosmic neighborhood), quantum particles perceive radiation with respect to **non-unitary**-spacetime (non-single-valued time metric and non-single-valued space metric, which comprise the quantum particles' wave nature). In such case, since an electron's essence incorporates a sinusoidal range of values of the time metric and the space metric, this gets the electron perceive a classical EMW of certain classical frequency as a superposition of alternations of higher and of lower frequencies of diminishing strengths. Such type of superposition can be described by **Fourier** transform and is responsible for a beat, or **quantum** of energy, referring to the Planck constant h, whose content measures on the basis of imaginary units of time, as described in section 7. The multiplication of h with the frequency v of the wave yields the energy of the photon, as per the relation $E=hv$. (Notice that it symmetrically holds for the quantization of charge and the associated energy $W=Qk$ as described in section 8).

According to the theory of the electroweak interaction, the photon is described by the relation

$$\gamma = \cos\theta_w\ B + \sin\theta_w\ W_3 \tag{9-1}$$

This expression should be understood to reflect on how each field component of the photon is perceived with respect to the reference frame of the interacting elementary particle (fermion), in non-unitary-spacetime. The angle θ_w corresponds to the Weinberg angle, which identifies to the phase shift between the space and time curves in figure 1-1. This angle does not refer to capacitive units (Farad) included within the permittivity term nor the inductive units (Henry) included within the permeability term of an equation like (5-7). The involvement of the Weinberg angle signifies that the interaction of the EMW with a fermion engages in an orthogonal interaction which refers to a phase shift in space, instead of a phase shift in time. The specific angle θ_w dictates how the wave's displacement electric and displacement magnetic fields (described through section 5.3 point 2) interact with elementary particles' waviness of imaginary space and imaginary time, allowing them to change the energy state of an electron involved. That the expression (9-1) involves a cosine component and a sine component, indicates that these two components should also be orthogonal to each other, where one of them should reflect the phase shift as it is perceived from the viewpoint of time, and the other should reflect the phase shift as it is perceived from the viewpoint of space. And yet, the reference of θ_w to space and time seems to be complementary with respect to the one of electronics.

In an effort to approximate the association (and compliance) to Maxwell's equations which describe the propagation of EMWs, we shall copy here the simplified versions of those, as follows

$$dE/dx=dB/dt \tag{5-6) or (9-2}$$

$$(1/\mu_o)\ dB/dx = (\varepsilon_o)\ dE/dt \tag{5-7) or (9-3}$$

While in a conventional sense these two relations are considered complementary to each other, in section 5.3 we presented how (9-2) may serve as a limiting case of (9-3), when the values of capacitance C and inductance L

contained within the regulators ε_o and μ_o are in balance to each other and the wave is travelling at the speed of c. In such limiting case the propagation is not subject to impedance, and that condition resembles a travelling EMW.

The basic relation (9-3) may be expressed more compactly in terms of **displacement fields**, where $H=(1/\mu_o)B$ is the displacement magnetic field and $D=(\varepsilon_o)E$ is the displacement electric field, and these two relate to each other as follows

$$dH/dx = dD/dt \qquad\qquad\qquad\qquad \text{(5-17) or (9-4)}$$

Actually, since D involves ε which in turn involves the capacitance C, and capacitance may be a complex entity (involving a Real as well as an imaginary part), D is also a complex entity, which refers to

$$D = \varepsilon\,(E\cos\theta - i\,E\sin\theta) \qquad\qquad\qquad \text{(6-29) or (9-5)}$$

In the latter relation, the Cosine coordinate refers to the involvement of the Real coordinate of the electric field $E\cos\theta$ in effects of action, while the Sine coordinate refers to the involvement of the Imaginary coordinate of the electric field $iE\sin\theta = \bar{E}\sin\theta$ in effects of reaction (involving the electromotive potential which associates to the displacement of charge). Furthermore, it should similarly hold for the displacement magnetic field, as follows. Notice that in this relation, the Sine coordinate $iB\sin\theta = \bar{B}\sin\theta$ should refer to the inverse energy version of the magnetic field, with radial-like field lines.

$$H= (1/\mu)\,(-\,i\,B\sin\theta + B\cos\theta) \qquad\qquad\qquad \text{(9-6)}$$

As per the above, the relation (9-1) appears to refer to the interaction of an atomic electron with the *real* coordinate of the displacement version of the electric field D whose raw units were previously shown to resemble the raw units of the magnetic field, and that may relate to the component "$\cos\theta_w\ B$" as perceived from a fermion's reference frame. Likewise, the interaction with the *imaginary* coordinate of the displacement magnetic field H whose raw units were shown to resemble the raw units of the electric field, may relate to the component "$\sin\theta_w\ W_3$" as perceived from a fermion's reference frame. In such case, the field components of an EMW that interact with an electron seem to involve displacement of fields, which both have radial-like field lines. While these are considered to propagate in space in the form of an EMW with velocity c, this propagation seems to concern "repetitive displacement shifts" instead of "motion", as described in section 5.3. As the fields alternate between each other they seem to displace in time in progressive steps; one of the fields shifts in time capacitively C/dx, then the other shifts inductively L/dx, where these shifts refer to the regulators ε and μ. The particular phase shifts $+\Delta t$ and $-\Delta t$ do not take *any* degree of angle (like it holds for L-C electronic circuits depending on the properties of the circuit elements), since an electromagnetic wave does not associate to any inductive or capacitive element of any circuit. The value of the phase shift is fixed and refers to the Weinberg angle. But while an inductive or capacitive phase shift in electronic circuits develops in time, the Weinberg angle was explained to concern a shift *in space*. That may be tackled through an issue of an orthogonality: **Just like an imaginary electric coordinate may transmit displacement-current I_d, the imaginary magnetic coordinate may transmit the equivalent of "displacement-negative-momentum p_d" in** *space*. That the magnetic coordinate may engage in that, may be figured through the raw units of the magnetic field which were previously discussed to associate to $(\sec^2/m^3)$, where that may be re-written in the form $(\sec^2/m)(1/m^2)$, or expressed in the form $(dt^2/dx)(1/dx^2)$. The later includes two terms; an acceleration-like term (dt^2/dx), and a flux-measuring term $(1/dx^2)$. The former resembles the negative reciprocal of the units of the gravitational field, since $d(1/x)/d(1/t)^2=(dt^2/dx)$, while the latter describes a flux (actually a "flow", since it takes place with respect to positive space, and not with respect to negative-reciprocal space). And this flow resembles displacement of momentum, which is why a photon may change the energy state of an electron, which is (in terms of energy) matter-related, not charge-related. However, since this displacement momentum p_d carried by the photon is negative (thus imaginary with respect to the real world), the photon appears as if it were a massless particle.

Furthermore, when the step-by-step displacement shift interacts with the electron in non-unitary-spacetime, it is perceived as a superposition of alternations of higher and of lower frequencies of diminishing strengths, which causes the interaction behave as a lump of energy in accordance to the Fourier transform, and that brings an interaction where the exchange of energy refers to the photon γ. In loose words, while an interaction of an EMW in a classical physics context concerns oscillatory values of B and E and fixed values of inductance L and capacitance C, quantum particles concern the opposite; an interaction with oscillating values of L and C as per the particle's quantum nature, which comes along with fixed quantized values of the particle's magnetic moment, and electric field E. In this case, the absorption of a photon may add to the existing balance of C and L in the electron's formation and cause a change in energy state and a corresponding change in quantum numbers attributed to the energy of the field components of the γ, and vice versa for the case of emission of a photon. A similar interaction in opposite sign of energy appears to correspond to the Z^0 boson as shall be considered right next.

The Z^0 boson

The Z^0 is an electrically neutral particle, and is considered to be one of the mediators of the weak interaction. According to textbook theory, it is described through the following relation:

$$Z^0 = \cos\theta_w\, W_3 - \sin\theta_w\, B \qquad\qquad (9\text{-}7)$$

As previously, this relation gets to be structured on the basis of how the interaction of this particle is perceived from the reference point of the fermion involved, and not from the point of view of the classical physics (macroscopic) world. Here the B component comes through a Sine term (instead of a Cosine term which was the case for the photon), and the W_3 component comes through a Cosine term (instead of Sine as was the case for the photon). This inversion suggests that **the Z^0 corresponds to a negative-energy cousin to the photon**. Since the field components involved concern opposite energy, they should interact with the imaginary part of an electron (seemingly concerning the equation of Sines (8-7)). Lisewise, this particle should ordinarily couple with leptons of opposite energy, like neutrinos and antineutrinos residing in the cosmic region of their ordinary habitat (in symmetry to as a photon interacts with an electron in our own cosmic region).

That the Z^0 particle's field components interact in opposite energy, signifies that these field components (and hence this whole particle) should quickly become **evanescent** in the classical part of the universe, same as it generally holds for waves of negative energy, which become evanescent. This is actually the reason behind the very short range of interaction of the Z^0 (concerning small subatomic distances) and its corresponding very short lifetime. Notice that the attribution of its short lifetime to the evanescence of its fields constitutes a complementary approach to the conventional notion that the short range of action of this particle is due to its high mass (in conjunction with the uncertainty principle). According to the new approach, the short-range of interaction of this particle is of similar nature to the short-range extend of transmission of a wave in frustrated total internal reflection, or to the short range of displacement of current through a capacitor's plates, or to the short extend of matter wave penetration (and corresponding displacement-momentum) during barrier penetration. What differentiates its bosonic (particle-like) behavior to these wave-born effects, is that its interaction with fermions takes place in non-unitary spacetime, so the Fourier transformation provides to Z^0 its particle-like nature (similar to as was explained to hold for the photon). Therefore, **the Z^0 boson should be understood to resemble an interaction of an evanescent EMW in non-unitary spacetime**. Actually, for similar reason to as the displacement of gravitational flux (displacement of the particle's matter wave) through a barrier concerns a displacement-momentum p_d, and the displacement of electric flux through the plates of a capacitor translates to a displacement-current I_p, it similarly applies for the Z^0 particle. Here, the associated displacement fields' evanescent component resembles transfer of **displacement-momentum**, as if the particle supposedly carried mass (which it doesn't exactly do). More particularly, the displacement-momentum in this case should not be attributed to an inertial-or-gravitational related effect, but to an **anti-magnetic** (which we could also call "weak-magnetic") effect instead. That is, the $-\sin\theta_w B$ coordinate seems to refer to the displacement of antimagnetic field in space, which applies in the form of an anti-magneto-motive effect as that is perceived from a fermion's point of reference. It is an inductive process, which displaces the analogy of "opposite momentum to the other end", for reasons of conservation of momentum. It is a reaction effect, of negative energy, and therefore applies through a tunneling process. While the corresponding impedance makes the Z^0 appear as if it carried mass, the process could rather be considered to concern displacement of **"pseudomass"**, or more accurately **"pseudo-momentum"** in the sense that it involves the transmission of "displacement momentum", the mechanical counterpart of a "displacement current"}. **That this displacement momentum is of positive energy gets the Z^0 apper as if it carried mass, while in the case of the photon the displacement momentum was of negative energy (thus imaginary to the real world), hence the photon appears being massless.** The notion that the Z^0 does not carry conventional mass may reach support via a comparison between the amount of this particle's "pseudomass" and the amount of mass carried by the weak interaction bosons W^+ and W^- which shall be considered ahead. The transfer of displacement-momentum is of temporal character, lasting only for as long as a transitional interaction and corresponding displacement takes place, similar to as it holds with the displacement current in electronics. The difference here is that the Z^0 interacts in non-unitary spacetime (as a superposition of alternations of higher and lower frequencies of diminishing strength), so it exhibits a particle-like nature. Furthermore, aside its short range of interaction, the mediated displacement of negative momentum should be taking place instantaneously, since it concerns a loop-like phase shift (just like all elements of a bicycle's wheel rotate concurrently, instantaneously). The property of instantaneity is also fine quantitatively, since the field components of Z^0 are of negative energy. That this reaction is mediated instantaneously can also be traced in the fact that the interaction involving a Z^0 boson exchange is "**elastic**", meaning that the kinetic energy is conserved. In fact, the "elastic" interaction attribute is a feature of the electromagnetic interaction, and not of the weak interaction, which is another clue in supporting the notion that the Z^0 **actually concerns the electromagnetic interaction** (not the weak interaction as conventional theory assumes).

In terms of units, that this particle may serve as a carrier of negative momentum (pseudomass), could be reflected in the units of the magnetic field (sec^2/m^3). Same as presented above for the photon, as well as it was mentioned in section 5.3, these units could be re-written by splitting the original term to (sec^2/m)($1/m^2$) which may also be expressed in the form $(dt^2/dx)(1/dx^2)$. That includes two terms; an acceleration-like term (dt^2/dx) as well as a flux-like term $(1/dx^2)$. The former resembles a negative reciprocation of the units of the gravitational field, since $d(1/x)/d(1/t)^2=(dt^2/dx)$, while the latter describes a flux, which is literally a "flow" since it develops per space (not reciprocal space). In this case, the negative reciprocal of the units of the gravitational field accounts as an imaginary gravitational field, and such imaginary character gets it resemble displacement of momentum, in analogy to as the imaginary electric field variant may account for displacement of charge. In this sense the antimagnetic constituent of the Z^0 gets to account for the transfer displacement-momentum.

Regarding the $cos\theta_w W_3$ component of the Z^0 particle, that seems to reflect the displacement anti-electric field (the negative energy version of the displacement electric field, just like the one that mediates the flux of displacement current in-between the plates of a capacitor), which may engage in transitional cases due to the existence of the phase shift between the space and time metrics. Only here, the interaction takes place in non-unitary spacetime. During its interaction, the Z^0 displaces anti-charge (the imaginary version of charge, which may also be referred to as "weak interaction charge". **As the anticharge is imaginary to the real world, the Z^0 appears to be neutral, hence it neither exhibits a magnetic moment. However, as anticharge is the ordinary charge of neutrinos and antineutrinos, Z^0 ordinarily couples with these particles**, which allows it to mediate interactions involving a decay to a fermion and its opposite energy counterpart, for example an electron and an antineutrino. In such cases the exchange of anticharge is mediated without noticing (since anticharge is imaginary with respect to the real world) so the Z^0 appears neutral to the classical world. In that sense, the Z^0 may be loosely considered a quantized mediator of a "displacement-antielectro-antimagnetic" wave, in a similar sense to as the photon may be considered a quantized mediator of a displacement-electro-magnetic wave.

Furthermore, that the Z^0 involves anti-field components, may well suggest that its field components may also operate in conditions of opposite energy (the space and time metrics having opposite sign), in which case its description may also be express without permittivity and permeability terms, in a similar sense to as (9-2) does not need permittivity and permeability terms, while (9-3) does require these terms). Furthermore, the Z^0 should be a long-lasting particle in negative energy. By extend, in the the negative space metric it should interact with neutrinos and antineutrinos just like the photon interacts with electrons.

The W^+ and W^- bosons

Regarding the W^+ and W^- weak interaction's bosons, these are described through the following relations

$$W^+ = (W_1 - iW_2) / \sqrt{2} \tag{9-8}$$

$$W^- = (W_1 + iW_2) / \sqrt{2} \tag{9-9}$$

The field components of W_1 and W_2 should be orthogonal to the field components of the γ and Z^0 in the sense that they should involve a swapping between the parameters of space and time, gettings the Ws to relate to fields of mechanics instead of electronics. This may be supported by the fact that the mass of the $W^\pm$ (denoted as M_W) is lesser to the mass of the Z^0 (denoted as M_Z) by a coefficient of $\cos\theta_w$, where θ_w is the Weinberg angle, as per the relation (9-10). That relation may imply that **the mass of the $W^\pm$ arises through the conventional Higgs mechanism (having inertial impedance caused due to the Weinberg phase angle between the space and time metrics) and therefore associates to an imaginary coordinate of energy which is spent without doing work in the real domain**. Notice that since the Ws are negative energy particles, the coordinate used in (9-10) is a *cosine* coordinate, instead of a sine coordinate which would apply for matter of positive sign. Also, their inertial properties are fundamentally different to those of the Z^0 (which was stated to concern displacement of pseudomass).

$$M_W = M_Z \cos\theta_w \tag{9-10}$$

This relation also suggests that the field components of the $W^\pm$ already incorporate a coefficient (Cosine or Sine) with respect to space, which is why the relations (9-8) and (9-9) between W_1 and W_2 only need an imaginary term i in one of the two field components, and do not need Cosine or Sine terms in addition to that. Here, the imaginary term indicates that the two components are based in opposite energies, so they are orthogonal to each other in a similar sense to as the displacement electric and displacement magnetic fields were stated to be orthogonal to each other in electromagnetism. In that case, **the imaginary term i next to W_2 is there for same reason that the magnetic permeability is reciprocated ($1/\mu$) with respect to the electric permittivity (ε) in electronics**.

The involvement of a phase angle θ_w in (9-10) also indicates that the component fields should have a displacement-field character. In particular, the W_1 seems to reflect the transfer of momentum, and -given its reaction nature- it seems to associate to a displacement *gravito-motive* field (the field that was described in section 6.2 to mediate the inertial reaction force). The inertial behavior of this field seems to be degenerate in terms of direction of time, and this holds for the same reason that both matter and antimatter were explained to be gravitationally attractive (the reason being that to the parameter of time in their units is squared). Therefore, the W^+ and the W^- seem to actually involve W_1 and $\overline{W}_1$ (instead of W_1 only), displacing "opposite momentum" and "opposite antimomentum" correspondingly, where both of those are of same sign of matter, and gives to both these particles the characteristic of carrying inertial mass.

In the other hand, the term $+iW_2$ or $-iW_2$ seems to reflect the transfer of charge (positive or negative), and that seems to correspond to a displacement *gravito-magneto-motive* field (this being a negative energy counterpart of the displacement gravitomagnetic field). **The involvement of this field resembles displacement of charge, however it actually displaces pseudo-charge**, very much like the Z^0 resembles carrying mass while it actually displaces pseudo-mass. The W_2 can therefore be understood to be a mediator of this imaginary field. The interaction with fermions (where the latter were explained to live in non-unitary metrics of space and time) gets the Ws behave as particles, same as it was explained to hold for all other bosons. And **since their fields have imaginary character with respect to the classical world, they become quickly evanescent so these particles have very short range of action**.

The Graviton

If we consider that the field components of the W^+ and W^- bosons associate to the displacement versions of the gravitational and gravitomagnetic fields, then it is of reason to look for association between these particles and the hypothetical particle called **graviton**, which is assumed to serve as a gravitational cousin of the photon.

Gravitational waves are currently treated differently to electromagnetic waves due to two basic assumptions:
(a) matter is conventionally considered to come in only one "sign" (always gravitationally attractive), while charge comes in two signs, positive and negative, and
(b) gravitational waves associate to an inertial oscillation, while electromagnetic waves do not relate to inertia.

However, the model presented here allows for a gravitational oscillation to take place in symmetry to the electromagnetic, as it assumes for the existence of positive and negative matter (as in electrons and in neutrinos), with both of which the $W^{\pm}$ bosons couple. As for the attribute of inertia, electromagnetic effects are subject to the symmetric effect of inductive impedance, in fact, inductance associates to a phase shift in time Δt much like inertia was explained to associate to a phase shift in space Δx. And as the field of gravity corresponds to a force-gradient field, the gravitational waves arise from an alternating acceleration of matter (as an alternation in momentum), much like EMWs arise from an oscillation of charge (e.g. an alternation in current).

In that sense, gravitomagnetic waves should be possible to treat in a similar way with electromagnetic waves, subject to the swapping of roles between the parameters of space and time (as it was stated to apply between the gravitomagnetic and electromagnetic interactions in general). We shall therefore refer to conventional equations describing the propagation of EMWs. For reasons of simplicity we shall address the relation (9-11), leaving aside, for now, the complementary relation (5-7) which involves the permittivity and permeability terms. If we express the gravitational wave propagation under a symmetric expression, which would look like (9-12), that would be wrong as it misses to take into consideration the swapping between the denominators dt and dx. What seems to hold, is that if in the electromagnetic interaction a field (e.g. the electric field) changes with respect to space dx, then in the gravitomagnetic interaction a corresponding field (e.g. the gravitational field) should change with respect to the negative reciprocal of space $-(1/dx)$, and it similarly holds with respect to time. This brings the aforementioned swapping between the parameters of space and time in the denominators, consequently the propagation of gravitomagnetic waves should rather be described by a relation of the form (9-13). And if this is so, that would signify that the existing theoretical assumptions regarding the nature of the graviton may not rely on proper ground.

$$dE/dx = dB/dt \hspace{4cm} \text{(5-6) or (9-11)}$$

$$dg/dx = dT/dt \quad \text{[WRONG]} \hspace{3cm} \text{(9-12)}$$

$$dg/\boldsymbol{dt} = dT/\boldsymbol{dx} \hspace{4cm} \text{(5-35) or (9-13)}$$

$$dg_d/dx = dT_d/dt \hspace{4cm} \text{(9-14)}$$

Recall that in accordance to material presented in section 5, the relation (9-13) is only partially correct, as it corresponds to a limiting case of a more general relation which involves the corresponding negative-energy displacement versions of the fields $g_d=(\varepsilon_g)dg/dx$ and $T_d=-(1/\mu_g)dT/dt$. In that case, the involvement of the mechano-permittivity $\varepsilon_g=\Delta x/dt$ and mechano-permeability $1/\mu_g=-dt/\Delta x$ (within the displacement field versions) causes an effective re-swapping of the parameters of time and space, making the denominators in (9-14) turn back to look similar to the ones of (9-12). But still that does not correct the issue that the existing theories about the gravitons seem to rely on wrong ground. Actually, the inverse use of dt and dx in the denominators seems to apply as equivalent to having missed to affect a reciprocation of parameters, and that results to a calculational fault that assumes that the hypothetical graviton has a spin equal to the reciprocal of ½ (instead of ½), where that makes for spin 2 (since 1/(½)=2).

In accordance to the above it seems that the characteristics of the hypothetical graviton and its corresponding antiparticle (speaking with regard to our own cosmic region) should rather concern the W⁺ and W⁻ bosons. And since these particles' field components are of negative energy, they quickly become evanescent in the real world, so these particles have very short range of action.

Bosons vs. Fermions

In relation to material presented in the previous sections, the difference between classical oscillations and quantum oscillations is that the later concern oscillation in non-unitary space and time (that is, in oscillating time fastness and space stretchiness). That oscillation gets external waves of a certain frequency being perceived as superpositions of alternations of higher and lower frequencies of diminishing strengths. Such type of superposition can be described by Fourier transform, and brings the associated "quantum" of energy corresponding to the Planck constant. While the non-unitarity directly concerns the essence of fermions, the interaction of fermions with classical waves gets the latter behave as packets of energy, having particle-like characteristics which correspond to the bosons. More particularly:

- Bosons concern interactions of the waviness of displacement E and B fields (of positive or negative energy field versions), and likewise of displacement g and T fields (of positive or negative energy field versions), as these fields are perceived from the point of view of fermionic particles. The action of these fields is counterbalanced through fixed values of C or L, and C_g or L_g, where that allows for the involvement of corresponding reaction potentials, as the electromotive, the magnetomotive, the gravitomotive, and the gravitomagnetomotive.

-   Fermions, in turn, concern the waviness of C or L, and likewise of C_g or L_g (of positive or negative energy versions), which is counterbalanced through fixed values of E, B, and g, T fields (comprising these particle's fields).

Fermions seem to be subject to an additional interaction as well, which is orthogonal to both of the ones mentioned above. That is, fermions seem to involve a cross-fields interactions (seemingly between L_g and C) where that reflects on the coexistence of matter and charge in particles like the electrons. In such way, nature gets to balance energy in the space and time domains, in the form of particles. That actually makes it reasonable to consider the co-existence of matter and charge in what we may call a "**chargo-matter**" (or "electro-gravitational") interaction, of which the "mediators" are simply, the **fermions**. So, in a sense, an electron can be considered to carry charge and matter in similar sense to as a photon carries electric and magnetic fields.

Notice that in such way, energy is not only balanced between positive and negative sign (real and imaginary, with respect to the real world), it is also balanced between the space domain (for charge) and the time domain (for matter). In relation to that, out of a particle's total energy $E=mc^2$, one-half seems to concern energy in association to the particle's matter, and the other one-half seems to concern energy in association to charge (as approximated in section 8). And since these two forms of energy concern opposite domains (the time and space domains), they counterbalance each other. Furthermore, the split of energy between the real and the imaginary domains, is also reflected in the half integer value of particles' spin properties.

In the case of bosons, instead, the alternating field components seem to both project on the real domain (one field component associating to an oscillation in space and the other to an oscillation in time), or to both project to the imaginary domain. In the case of a photon, for instance, the electric field's action seems to concern curvature in imaginary space and the magnetic field action seems to concern curvature in imaginary time. But since one of them concerns space and the other concerns time, the magnetic field is orthogonal with respect to the electric field. Yet the magnetic field may couple with the imaginary part of a particle's wave function and yield real outcome, and in such way it engages that of in same sign of energy as the electric component. The opposite applies for the Z^0 mediator boson where both its component fields project in opposite (negative) energy. The same general concept seems to apply for the $W^\pm$ bosons, subject to a swapping of parameters of space and time. In this way, as a bosonic particle propagates, energy alternates in balance between the time and space domains, both at the same sign of energy (either both positive or both negative), hence the integer value of particle spin.

When numerous bosons co-exist in a region, or numerous fermions occupy a region, nature tends to maintain the total energy balanced out. When **bosons** are emitted or absorbed spontaneously, their non-coherent phase degrades the even-ness of energy balancing. To avoid a corresponding unbalance which has the form of noise, nature "prefers" to have bosons being emitted coherently in order to maintain the energy even-ness and balancing out. This takes place via favoring stimulated emission of photons in a coherent process which keeps the energy unbalance lower. This is what holds in the emission of laser light, which drives bosons to bunch up so as to occupy the same quantum state, with no restriction on the number of them joining.

An analogous bunching can be considered to hold for fermions with respect to waviness in the metric of time that constitutes their gravitational field of each particle. In this case, **particles tend to bunch up toward the region where they all have same phase in the waviness of the metric of time, where that corresponds to what we perceive as the gravitational interaction**. A somewhat similar feat seems to apply with respect of the electric field as well, however, since the electric field concerns the curvature of imaginary space (negative reciprocal space), the interaction of two charges projects to the real world as an interaction of two imaginary entities, where the condition $i^2 = -1$ applies. Due to this negative sign, the force points in the opposite direction, hence a coherency-driven attraction in this case should be understood to involve charges of opposite sign.

As per the above, Maxwell-like classical relations could be prepared to describe the coexistence of matter and charge within the particle, as well as mechanical spin and spin magnetic moment for electrons. And the same holds for other elementary particles too, for example neutrinos that reside in the cosmic region which constitutes their own stable habitat, where their negative sign of matter co-exists with anticharge.

Certain additional comments concerning quantization criteria of fermions and bosons are considered in the next section.

The Higgs boson

A note is provided here on the Higgs boson even though does not belong to the electroweak interaction. The reason is the particular association this particle has to both the electromagnetic and the weak interactions. As previously described, the Higgs field -which permeates spacetime- refers to the phase shift between the metrics of Space and Time, depicted in figure 1-1. The existence of this shift allows for leptons' opposite energy fields to engage in effects of reaction through their displacement-field variants. Furthermore, the particular value of this shift was introduced to refer to the Weinberg angle, where the exact value of this angle allows for the existence of quarks, and is responsible for the value of their charge, as shall be explained in section 10 ahead. The particle-nature of the Higgs field (behaving as a boson) arises the same way it does for all other bosons, which concerns how it is perceived from fermions' point of reference. As explained earlier, since a fermion involves a sinusoidal-like alternation in the fastness of passing time and the stretching of space, that gets any interaction of this fermion with external fields to take place as if the alternation consisted of a superposition of alternations of higher and lower frequencies of diminishing strength, and as per Fourier analysis that takes the form of a beat, or quantum of energy. That is what gets this field too to interact in the form of a boson. As the Higgs field is described to concern the phase shift between the metric of space and the metric of time, its nature should rather be stationary and uniform throughout the universe (at least over the cosmic range where stable matter of any form exists). Because of that, when this field interacts in the form of a boson, its spin value should be zero, as the theory predicts and experiments confirm.

10

Nuclear particles & Concepts on the Strong Interaction

So far, we have considered leptons, their involvement in the gravitational and electroweak interactions, and how these two interactions complement each other. What remains to be resolved is the essence of composite particles, in particular the nucleons, as well as to provide certain insights about the strong interaction which binds their constituents together. Nuclear particles (protons, neutrons) differ from leptons (electrons, neutrinos) in that they are composed of sub-particles, the quarks. These behave as quasi-particles, and as we shall figure, their particular behavior seems to relate to a phase shift, that gets them to partially dive into negative energy. These quasi-particles may reach stable form only if they combine with each other in a way that allows them to collectively satisfy basic quantization criteria, including quantization of charge.

We shall approximate the nature of these sub-particles step-by-step, starting with a comparison to the nature of the leptons, in particular the electron. In previous sections we have explained that an electron has a real part (of negative charge) and an imaginary part (of negative anticharge). Thus, we may consider the electron as a hypothetical dipole, of which only the one pole has Real essence, while the other "virtual pole" does not take real form and remains hidden, but may engage indirectly in effects of reaction thanks to the relation $i^2=-1$. The notion of such virtual dipole was illustrated schematically in the depiction of figure 5-5 which is reproduced here for convenience, in figure 10-1. The left part of this figure illustrates the concept of the electron dipole, depicted from a hypothetical "side view". As that figure shows, in real life we "view" the dipole from the upper part (see eye on top of the figure). From that point of view, we perceive only the real part of a field line (black arrows), and not the imaginary (hidden) part (grey arrows at the lower half of the figure. In the case of a free particle which is far away from other particles, we may consider the field lines to extend **radially** away from the point-particle toward the distant universe, so they appear to us as open-ended (same as in the depiction of figure 5-2 right part). Instead, from the point of view of the hypothetical "side view" each field line forms a hypothetical **loop** that reaches toward the distant universe and then turns imaginary (grey arrows) where that imaginary portion correspond to an anticharge field (which we may not directly view from "above").

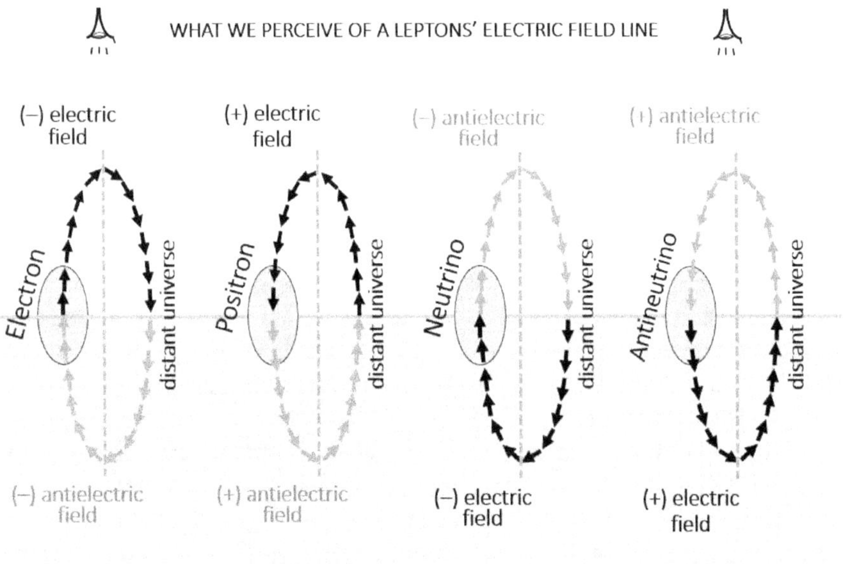

Figure 10-1: Re-print of figure 5-5, exhibiting the visible and non-visible (real and imaginary) electric field portions of the electron and the other leptons, as seen from a hypothetical "side view"

As also described earlier, from each field-line loop, only the portion <u>to the left of the vertical dashed line</u> refers to the particle, while the portion to the right of the dashed line refers to the other, virtual pole (nodal point or surface) at the distant universe.

It similarly holds for the positron (second from left in figure 10-1) of which we may directly perceive the electric field of the positive charge, but not the anti-electric field of anticharge. (The direction of arrows is only schematic, just to depict the notion of opposite direction of time that holds for this particle). It similarly applies for the neutrino (third from left), which exhibits to us its anticharge which however may not be perceived since it is imaginary with respect to our point of view, while that particle hides actual charge at its hidden part (that applies as a real field at cosmic regions of negative space metric). And it similarly holds for the antineutrino as well (far at the right).

Under the same concept, it similarly holds for lepton's matter as well, with the electron exhibiting to us its Real part of **positive** matter, which comes along with the conventional gravitational field lines, while there is also an imaginary (hidden) part of negative matter and a corresponding imaginary negative gravitational field. And here again, there is a hypothetical vertical dashed line, where the elementary particle's field refers to the left side of it. And likewise, it applies for the positron as well, having a real part of **positive** antimatter, and a hidden part of negative antimatter. However, a positron may not survive for long as a free particle in our cosmic neighborhood (the celestial region d'-b of figure 1-1) as its natural habitat concerns the negative metric of time, while the positive direction of time that prevails in our cosmic neighborhood makes this particle vulnerable to engage in an interaction with equal energy in positive time, leading to annihilation. Furthermore, the inverse holds for neutrinos and antineutrinos, which exhibit to us their negative energy part (negative matter or negative antimatter correspondingly), which is imaginary to us and that makes these particles behave as being massless in our cosmic region.

Just like in the theory of electric circuits the oscillations of Voltage and Current may be subject to a phase difference, an analogous phase shift applies between the waviness of space and time as illustrated in figure 1-1. That particular phase shift allows for a differentiated particle formation, whose matter wave is subject to a phase shift between the parameters of time and space. When such a phase shift applies, the particle is no longer behaving as a lepton, and it turns out that it may rather obtain the properties of a quark. In fact, not any angle of phase shift can allow formation of a stable particle of this type. It seems that there exist particular angles that may produce stable formations, if certain conditions of quantization are met. Such a phase shift of significance, concerns the angle of 30° (one third of the orthogonal 90°), which equals the Weinberg angle, which ascribes to the phase shift between the metrics, depicted in figure 1-1. If such an angle of phase shift is affected on the particle's oscillation of capacitance, **that should go along with a same degree of phase shift in the deployment of a fermion's electric field** radially on space. (Notice an analogy to the V-I phase shift in electric circuit loop oscillations, but now developing on space, and in the radial direction). That radial-like displacement of the fermion's electric field is fixed (not oscillatory), and gets a corresponding portion of the electric field to become imaginary, and concurrently a part of a particle's imaginary part (corresponding to anticharge) to displace toward the real side of the world. **This lets the fermion exhibit a phase-shifted electric field (and charge), so only a portion of the electric field (and charge) is let to project on the real world, while the rest has become imaginary. In this case the particle does not carry the attributes of a conventional lepton anymore, instead it corresponds to a quark.** This field shift is more apparent in radial-like fields where the field "arrows" shift radially outward or inward, so this corresponds to a fixed displacement in the field's flux (as that deploys radially over space). The corresponding concept of the shift of the electric field (a fixed radial displacement in flux) is illustrated in figure 10-2, which is an extended version of figure 5-18 as it has more information added into it. Note that the same basic concept described here for the electric field, applies for the gravitational field as well.

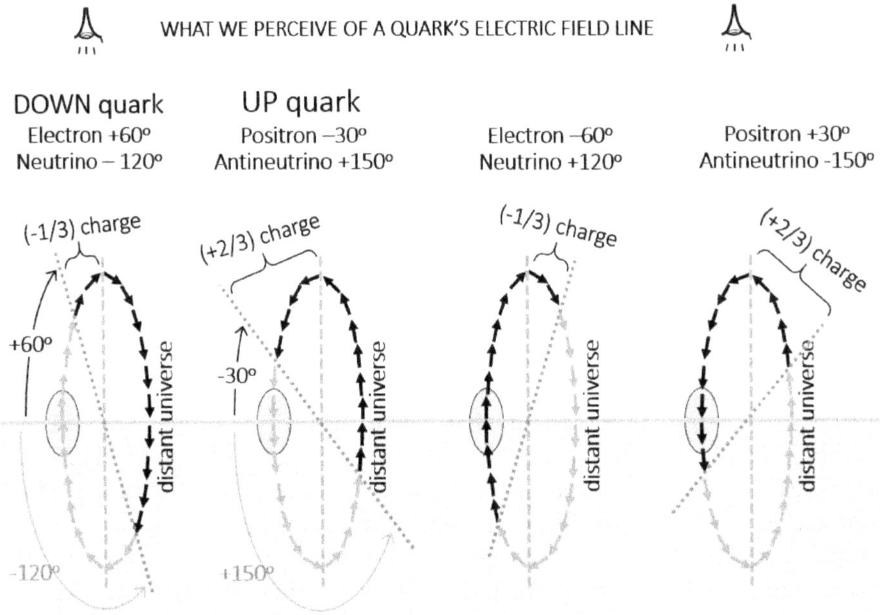

Figure 10-2: A phase shift of a particles' electric field in the direction of the field arrows by ±30° or ±60° transforms a lepton to a quark

Remember, that the field portion which ascribes to a particle concerns the part *at the left* of the vertical dashed line. In the figure's left depiction, one may notice that the "negative electric field" arrows of an **electron** (black arrows) have been phase-shifted clockwise (forward) by (90°-30°=) 60°, leaving only 1/3 of the real electric field active (the black arrows at the upper-left quadrant) ascribing to a value of charge $Q = -1/3$ e. (This is so, since only 1/3 of the black lines are visible from "above" which represents the "real side" of the world, while the remaining grey arrows to the left of the vertical dashed line ascribe to an imaginary field version so they do not constitute ordinary charge with respect to our perception). Notice also, that the same formation arises by a counterclockwise (backward) shift of a **neutrino** by -120°. The resulting particle formation in any such case corresponds to 1/3 of an electron (black arrows), and 2/3 of a neutrino (grey arrows at the upper left quadrant). And as per its net charge of value -1/3 e, this corresponds to a **down quark**.

Likewise, in the second-from-left depiction, one may notice that the "positive electric field" arrows of a **positron** have been phase shifted clockwise (backwards) by -30°, leaving only 2/3 of the field active (black arrows at the upper-left quadrant) ascribing to a value of charge $Q = +2/3$ e. The rest of the arrows (grey arrows at that quadrant) correspond to anticharge (which is imaginary). Notice that the same formation may also arise by a counterclockwise (forward) shift of an **antineutrino** by +150°. The resulting particle formation in these cases corresponds to 2/3 of a positron (black arrows), and 1/3 of an antineutrino (grey arrows). And based on its net charge of value +2/3, it corresponds to an **up quark**.

In relation to the above, alternative formations of same value of net charge may also exist (either -1/3 or +2/3), as shown in the last two depictions at the right of the figure 10-2. A point of attention with regard to these particular formations, is that the partial real electric field (black arrows) seems to lie at the upper-*right* quadrant (which is to-the-right of the vertical dashed hypothetical line). That seems to ascribe to the field of the "other pole" (toward the distant universe), so in a first reading such formations may not constitute a conventional quark. However, such formation could still interact in effects of opposite energy, as in relation to the field portions at the lower-left quadrant of their depiction. That part might actually engage in the strong interaction, where that angle of shift might actually play a key role in the confinement of quarks into stable composite particle formations.

Similar type of phase shifts may allow for the existence of quarks of opposite charge, which would therefore be referred to as "antiquarks", as illustrated in figure 10-3. For instance, a positron whose electric field is shifted backward (clockwise) by 60° or an antineutrino shifted forward (counterclockwise) by

120° would create a particle of charge equal to +1/3 e, as shown in the depiction first-from-left of figure 10-3. Likewise, an electron whose electric field is shifted forward (clockwise) by 30° or a neutrino shifted backward (counterclockwise) by 150° would create a particle of charge equal to -2/3 e, as shown in the depiction second-from-left of the same figure.

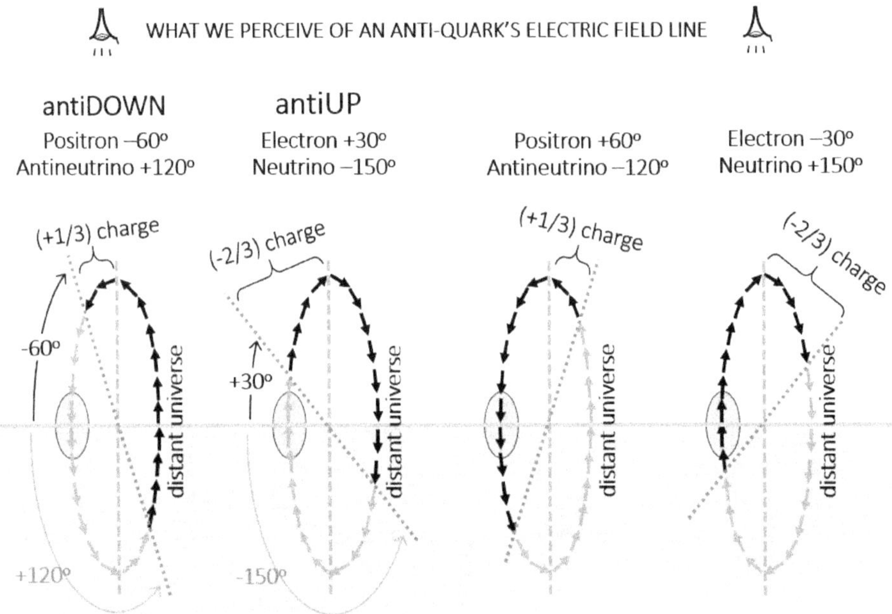

Figure 10-3: Schematic representation of antiquarks, shown from a hypothetical "side view"

Here again, alternative formations of same value of net charge may also exist (either +1/3 or -2/3), as shown in the last two depictions at the right of the figure 10-3. But in these formations the partial real electric field (black arrows) seems to lie at the upper-right quadrant (which is to-the-right of the vertical dashed hypothetical line). So that again seems to ascribe to the field of the "other pole" (toward the distant universe) thus in a first reading such formations may not constitute conventional quarks, however such attribute could still have them interact in effects of opposite energy through the field portions at the lower-left quadrant of their depiction, via the strong interaction.

The feasibility of existence of every quark variant is subject to a number of criteria. Recall that an electric field "line" relates to a curvature in the imaginary space metric, a gravitational field line relates to a curvature in the time metric, and in relation to what was considered through figure 6.4 the curvatures of time and space are subject to a phase shift between them, which refers to the Weinberg angle (or 90° minus the Weinberg angle) and rules the ways the particle may interact in effects of negative energy, which concern all interactions. What this tells, is that the space curvature nodes d and b (shown in figure 6-4) are "interrupted" by the change in sign of time curvature at the nodes d' or b'. And since only one of the later nodes may lie in-between the former, this may limit the ability of certain quark version to mediate the weak and strong interactions, and in relation to that, it limits the number of feasible quark variants into half. It seems that this limitation applies with respect to the allowed phase shifts concerning the electric field. At the same time, for the allowed quark variants, the phase shift with respect to the gravitational field seems to constitute the reason why gluons appear as if they contribute to a composite particle's mass, while they should actually possess pseudomass. (It applies under similar principle to as in the previous section the Z^0 boson was referred to appear as if it carries mass while it actually carries pseudomass).

The fields of individual quarks combine in forming the collective fields of the composite particle that they create (a baryon). This allows for quarks to meet certain criteria of quantization *as a collection of sub-particles*, while they would not be able to meet these criteria as independent sub-particles on their own. Through this way they may reach stable formation as composite particles, while at the same time this concerns the reason why quarks may not be found isolated in nature. Criteria that quarks need to meet collectively in order to constitute **stable** composite particles are similar to the criteria needed for leptons. Such criteria have as follows:

(a) Meeting quantization requirements
Quantization requirements that quarks need to meet as a collection of sub-particles, concern both its matter as well as its charge. To approximate what holds, we shall first make a reference to what applies for leptons:

In terms of charge (and anticharge alike), the quantized value of $\pm 1e$ arises as charge relates to a sinusoidal waviness in negative-reciprocal space as considered in section 8, and the specific value should come up via Fourier transform applied on space in similar way to as was explained in section 7 to hold for the quantization of h via a Fourier transform with regard to time. The reason that charge's quantized value for an electron corresponds to an *integer* value of $\pm 1e$ while matter's quantized lowest energy state has the non-integer value of $\frac{1}{2} h$, concerns the existence of the phase shift between the metrics of

Space and Time. Considering that: (i) matter refers to an oscillation in the parameter of time, (ii) the parameter of time needs to project upon space through the phase shift between the metrics of Space and Time, where the corresponding angle equals the Weinberg angle 30°, (iii) that projection should come through a Sine coordinate since Time is orthogonal and therefore imaginary with respect to Space, and (iv) it holds sin30°=½, that gets the quantized value of the minimum energy state of an electron to refer to ½ h. The electron charge, instead, refers to an oscillation in the parameter of space ("real space" if considered with respect to the electromagnetic interaction's point of reference) . Since it concerns an effect of space, it projection on space does not come through a Sine coordinate. Thus, any quantization of charge should not be accompanied by a coefficient like ½, it should come as a whole value, equal to the electron's charge -1e.

When it comes to quarks, these sub-particles can be regarded as leptons which are subject to an inherent 30° or 60° phase shift in fields, which only allows for a fraction of charge and a fraction of matter (and corresponding fields) to appear toward the real world (as illustrated over the last figures). In this case, while each quark may not match charge and matter quantization criteria by its own, nature provides a way to overcome this limitation when combining more than one quarks together, so that the combination of quarks may match the quantization of charge and of matter in a collective manner, as it holds in protons or in neutrons. Existing theory suggests that a **proton** is composed of two "up" quarks (symbol "u", each having charge +2/3) and one "down" quark (symbol "d", having charge -1/3), this making up a "uud" of 2/3+2/3-1/3=+1 charge for the proton, which now satisfies the charge quantization criterion. Likewise, a **neutron** is composed of one "up" quark and two "down" quarks, making up a "udd" of 2/3-1/3-1/3=0 charge for the neutron. It is also of particular interest, that in relation to the upper-left quadrants of the drawings on figures 10-2 and 10-3, each quark also carries anticharge. That is, **a proton (a udd particle) has anticharge of -2/3+1/3+1/3=0, while a neutron (a uud particle) is not exactly a neutral particle as it carries anticharge of -1/3+2/3 +2/3=+1ē**. And likewise, an antiproton has anticharge +2/3-1/3-1/3=0 and an anti-neutron has anticharge of +1/3-2/3-2/3=-1ē, while the anticharge of these nucleons also associate to corresponding antimagnetic fields, engaging in opposite-energy processes that leak energy toward the classical world. On a note in relation to that, while for leptons we needed 4 separate particles to account for all charge versions (namely the electron having negative charge, the positron having positive charge, the neutrino having negative anti-charge, and the anti-neutrino having positive anti-charge), for the case of baryons we only need the proton (having positive charge) and the anti-proton (having negative charge), but we do *not* exactly need a "negative proton" (since that is already embodied in the neutron which has charge +1ē), neither we exactly need a "negative anti-proton" (as that is already embodied in the anti-neutron, since that has charge $-1\bar{e}$). These above conditions are very significant. For one thing, they provide new clues on what holds inside neutron stars, where, neutrons inside these objects seem to play the role of atomic protons do it our classical world, protons play the role of neutrons, and neutrinos seem to allocate around their nuclei-like formations much like atomic electrons allocate around nuclei in our cosmic region. Even more importantly, new clues arise on the physics of conventional stars and planets: While the sum of the positive charge of all their protons almost equals the sum of negative charge of all their electrons, **the sum of the anticharge of all their neutrons make each celestial body an anticharge monopole. That, in fact, directly associates to the curvature of positive space around celestial bodies**, just like these bodies' matter directly associates to the curvature of positive time around them.

Notice that since a proton or a neutron has 3 quarks, which is not an even number, its net structure is not symmetric, so the composite particle exhibits a net spin ½ and behaves as a fermion. This is unlike what holds for mesons. A meson is made of a quark and an antiquark, its internal structure in terms of matter and charge balances out, so it behaves as a boson. Similar conditions to those of quarks apply for antiquarks, which have the same mass, mean lifetime, spin, but opposite charge and color-charge with respect to quarks. A point of interest concerns the fact that, since antiquarks are supposed to have opposite charge, one could argue that the up, charm, and top quarks are actually "antiparticles", since they carry "positive" charge (of +2/3). Positive charge is also carried by the antidown, antistrange, and antibottom, of value +1/3.

As per the above, composite particles should meet quantization requirements in terms of both the Planck constant and the value of charge. And each of the involved quarks behaves as 2/3 of a lepton of positive matter and 1/3 of a lepton of negative matter, or vice versa. Therefore, a quantitative description of the quantum oscillator concerning a quark, should be described by a linear combination of equations (7-8) and (7-9) with regard to matter (could use their reduced versions like (7-20)), as well as a linear combination of (8-8) and (8-9) with regard to charge, where all those seem to have an engagement of their own in the strong interaction.

(b) Overcoming the Pauli exclusion principle
The Pauli exclusion principle prohibits identical particles (fermions) from occupying the same energy state. This limitation extends to the contents of protons since they contain two "up" quarks and one "down", as well as the contents of neutrons since neutrons contain two "down" quarks and one "up". The two "up" quarks of the proton should be able to differentiate from each other in some way in order to not be identical, and the same applies for the two

"down" quarks of the neutron. According to conventional theory, such differentiation exists through the property of "color-charge", as each of the three quarks of a proton or a neutron has a different "color-charge". The color-charge comes in three values, which bear the symbolic names red-blue-green, but what each of these colors is meant to physically represent, is unknown. According to the new model, the existence of 3 color-charges seems to relate to the newly presented grounds, that the Weinberg angle corresponds to a third of $90°$, that a multiple of that angle may refer to a different quark, and more particularly that the shift may take place per either of the 3 dimensions of time which the present model assumes in the previous sections. The 3 dimensions of time in this case might refer to alternative states of phase shift of the quark's parameters of time, with respect to space. Actually, when an exchange in color-charge additionally involves a change in particle generation (to be discussed right next) then the triangularity of the three color-charges changes, and the cosine and sine coordinates involved in the displacement-field interactions no more simply concerns the Weinberg angle, but involves the combined Cabibbo angle. The alternation of color-charge between quarks may resemble the alternation of polarity between the plates of a capacitor, which was previously discussed to involve a displacement-potential. Each one of the color charge fields, named red-green-blue, comes in two opposite energies, for example the red comes in red r and anti-red $\bar{r}$, and that seems to correspond to a difference like the one between g and $\bar{g}$ in a gravitational equivalent, or T and $\bar{T}$. A link of color fields to the three dimensions of time would in fact associate the color charge to the gravitomagnetic interaction. Yet, for same reason as in previous sections the *raw* units of the magnetic field were considered suitable to mediate transmission of displacement-negative-momentum, here a similar symmetry may link color charge to displacement antimagnetic ($\bar{B}_\mathrm{d}$) properties.

(c) Avoiding annihilation
While electrons and positrons may be considered to live in positive energy, and neutrinos and antineutrinos to live in negative energy, quarks live in part in positive energy and in part in negative energy, concurrently. The positive energy portion is crucial for quarks to come into existence in the real world via combining with other quarks, while the negative energy part plays its own role by having its field components contribute in the strong interaction.

A point of consideration is why the protons (which have positive charge) are stable particles, while the positrons (which are also positively charged) and are truly elementary particles are not stable in our own cosmic region, as they quickly annihilate. The reason that this happens is that the positrons live under negative energy in time, so the cosmic region where they would be stable corresponds to d-d′ of figure 4-2 lower part. In that specific region the **product** of the energy in space and the energy in time is *negative*. In the contrary, in our own cosmic region d'-b the product of the energy in space and the energy in time is *positive*. That makes a positron vulnerable to match its energy with same energy of opposite sign (of an electron), and annihilate together. Quarks instead, which constitute the constituent of protons, partially engage in both positive and negative energy. This gets the fields from either sign of energy engage into interactions toward the opposite side of energy. (It's like having interactions relating to Maxwell's fourth law and Lenz's law taking place in parallel, with a probability, via the action of the partial "field coordinates" of quarks). Due to the corresponding opposite energy involved, the corresponding reaction fields' attributes are of very short range (as then they become evanescent). That prohibits quarks from interacting with quarks that lie far away (in neighboring composite particles), and limits their action to very close range, just between the quarks within the same composite particle. In this case, since the quarks of the particular composite particle have different values of color-charge (phase shifts), that does not let them match opposite energies and be able to annihilate. Consequently, that lets composite particles have extremely long half-lives.

(d) Binding of quarks through the strong interaction
In order to approximate certain behaviors of the interaction of color charge, we may indicatively refer to certain aspects of the electromagnetic (EM) interaction as a stepping-stone, as follows:

- The electromagnetic interaction involves two counterbalancing fields, the electric E and the magnetic B. In section 5.3 we considered how these apply through their displacement field versions, D and H.
- In section 6 we also considered that since the expressions of the displacement fields contain the electric permittivity and the magnetic permeability, which in turn involve capacitance and inductance that may be complex entities, the displacement fields may also involve a real as well as an imaginary component. This lets these field components become involved in different orthogonal effects.
- We also figured that interactions involving reaction potentials (like the electromotive or the magnetomotive) may get a change taking place in the real domain generate a change in the imaginary domain, and vice versa. For example, in Lenz's law a change in magnetic flux Φ_B (concerning a conventional magnetic field with loop-like field lines) may generate a change in the loop-like version of the electric field (corresponding to $\bar{E}$). And it similarly applies (in the opposite direction) in relation to Maxwell's fourth law where a change in electric flux Φ_E (concerning a conventional electric field with radial-like field lines) may generate a change in a radial-like version of the magnetic field (corresponding to $\bar{B}$).

- The above two types of interactions (electromotive and magnetomotive) concern displacement of energy toward different directions. The one may concern the displacement of charge (in the form of anticharge) between a capacitors' plates, and the other may concern the displacement of charge through a metal ring. One seems to concern the flow of energy from the real domain to the imaginary domain, and the other seems to concern the opposite, exchanging energy between differentials of time and space. We shall here loosely represent the former interaction as $\overline{D}H$, and the latter as $\overline{H}D$. Furthermore, while these processes concern interaction with fields of electrons, similar interactions may concern interaction with fields of particles of negative energy, like neutrinos.

In the case of quarks' interactions, the following differences seem to apply:

- Instead of displacement charge, we have displacement of phase-shifted fractional charge.
- Since one portion of the charge is of positive energy (ascribing to the electron-like portion of the quark) and another portion is of negative energy (ascribing to the neutrino-like portion of the same quark), this allows an interaction to take place in either direction. Loosely speaking, this is like having Maxwell's Law and Lenz's Law to apply in a combined way, with a probability for each. The result is an interaction that looks like $(\overline{H}D \pm \overline{D}H) / \sqrt{2}$, where the square root is for normalization purposes, signifying 50% chances for each part. But instead of D and H, in the present case the displacement fields involved concern the phase-shifted fractional charges.
- The co-involvement of the 30° phase angle and the 3 dimensions of time seems to allow for the interaction between the 3 quarks' fractional fields operate as of 3 interacting poles (instead of 2 poles in the electromagnetic case), ascribing to the 3 so-called "color-charges" of the strong interaction.
- Similar to as the fields of the electromagnetic (and the weak) interaction applies through the cosine and sine terms involving the Weinberg angle (approx. 30°), it should similarly hold for the strong interaction. A corresponding mechanism seems to allow for gluons' fields to apply in the form of displacement fields. The angle involved in gluon fields should rather differ from the Weinberg angle, due to the triangularity of the color-charge (having three poles instead of two). When an interaction that involves an exchange in color-charge additionally involves a change in particle generation (to be considered next), then the effective coordinate angle is also different to the

Weinberg angle, and concerns the combined Cabibbo angle of the weak interaction.

- In similar way to as in electronics we have an oscillation of an electric field generate an alternation of polarity between the plates of a capacitor (between plus or minus), in the case of quarks we have alternations between three modes (color charges), in a continuous exchange of so-called color-charge between quarks.

- Furthermore, similarly to as in non-unitary space and time the electromagnetic interaction exchanges energy in the form of photons, a corresponding interaction in non-unitary space and time as per Fourier transformation should get the color-charge interact particle-like in the form of gluons. In this sense the changing of colors between quarks get to correspond to an effective exchange of gluons.

- Each one of the color-charge fields, named red-green-blue, comes in two opposite energies, for example the red comes in red r and anti-red $\bar{r}$, much like the electric field comes in E and $\bar{E}$ (or in D and $\bar{D}$) as described earlier. Here, the negative energy version of a color field should actually be named "negative-color" (not "anti-color") charge.

- While each quark partially engages in both positive and negative energy concurrently, gluons should be thought of to carry both color and anti-color, interacting in a "combined state" with a probability. This may provide a match to current theory that a color state may -for instance- be $(r\bar{b} + b\bar{r})/\sqrt{2}$ where the square root is for normalization purposes. When a gluon is transferred between quarks, a color change occurs in both quarks. For example, if a red quark emits a red-antiblue gluon it becomes blue, and if a blue quark absorbs a red antiblue gluon it becomes red, as existing theory explains.

- The displacement field versions that comprise the essence of color-charge should have imaginary character, for similar reason to as capacitive and/or inductive phase shifts involve the imaginary version of the electric and magnetic fields (same as in the displacement of charge between a capacitor's plates which was explained to concern the imaginary electric field version). This is what gets the strong interaction to become evanescent very quickly, hence the strong force has a very short range.

- The different color charge of quarks of a single composite particle may affect on how it self-adapts with respect to other quarks, so they all constraint each other. That the alternative quarks of same net charge in figure (10-2) may affect their electric fields (arrows) at different ranges to the left or to the right of the vertical dashed line (where here the part to-the-right of the dashed line may refer to the field of the neighboring quark) may reflect on the gluon interaction being able to keep constant regardless the separation, letting color-fields behave as string-like objects called "flux tubes".

- Regarding the reason why baryons are heavy, this concerns the contribution of the gluons which bind the constituent quarks together. It seems that the motion of gluons is subject to an inertial-like potential which behaves in a similar way to as the Z^0 was considered to exhibit pseudomass in section 9. In this case, the involvement of all gluon variants (combinations of colors and anti-colors) add up to a large effective pseudomass, where that significantly contributes to the total inertial impedance, hence to the "effective mass" of a baryon.

<u>Particle's Generations</u>

Elementary particles may exist in higher generation versions, which exhibit substantially higher mass, while their electric and strong interactions remain unchanged. Higher generation versions exist both for leptons as well as for quarks (where quarks were explained above to correspond to phase-shifted leptons). Due to their corresponding high energy, the higher generation particles decay quickly so they are only found in extremely high-energy environments such as particle accelerators and cosmic rays.

The creation or decay of higher generation particles mainly involves the **weak interaction**, where the change in mass comes through the exchange of W$\pm$ bosons. The strong interaction, instead, refers to the change of color-charge through the exchange of gluons, where gluons bind the quarks together, and their fields affect inertial impedance in changing the motional status of baryons and this applies as a contribution to baryons' effective mass. Yet, the gluons themselves do not directly mediate change in mass during generation change.

It seems that the involvement of the W$\pm$s in the change of generation mimics an electromagnetic effect which seems to co-engage the photon γ and the Z^0. That could be described as per the following pattern of thought:

- The absorption or emission of a photon changes the electronic energy state of an electron or positron (but not the particle's charge).
- Likewise, the absorption or emission of a Z^0 should change the electronic energy state of a neutrino or an antineutrino (but not its anticharge), provided the neutrino or the antineutrino resides in the cosmic region of their habitat.
- While the photon does not carry charge, and the Z^0 should not carry anticharge, the Ws carry mass. That the photon does not carry charge, is because the photon's emission takes place *sideways* to the acceleration of an electron. (This is so, since with respect to that

direction the emission is radiative, it is not subject to a net inductive or capacitive impedance).

- In the direction toward the charge's acceleration, the emission is not radiative (except in relativistic conditions), but instead it may displace charge, like it holds between the plates of a capacitor. That displacement of charge has been discussed above to concern a wave effect of tunneling (despite the fact that it may resemble particle-like motion of charge). Even though the transfer of charge between the plates of a capacitor is a wave effect, it does not exactly concern the transfer of photons, since radiation is not plausible in that direction. An electromagnetic wave aiming at that direction is an evanescent wave (which dumps out quickly) since it has imaginary attributes. In this case, it is the imaginary essence of this modulation which mediates the transfer of charge, as described earlier.

- Notice that since a photon travels at c, time is not moving with respect to the photon, so the emission of a photon by an electron and its absorption by another electron may be considered an instantaneous process from the point of view of the photon. In effect, that resembles an effect of tunneling in time. The exchange of this photon between the emitting and the absorbing electrons results to a change in energy state and corresponding increase in energy of the receptor electron. The increase in the energy state of the receptor resembles an increase of energy in terms of matter (not in terms of charge). And this is because the photon was described to carry displacement-negative-momentum.

- If we consider a similar interaction in the orthogonal non-radiative direction between the plates of a capacitor, that could be assumed to involve the exchange of a virtual photon. (It is called virtual since it is considered to take place in the non-radiative direction). That interaction may be considered to involve the mediation of "displacement-current", instead of mediation of energy in terms of "displacement-negative-momentum" which was the case along the radiative direction.

- **If we were to describe the displacement of charge through the plates of a capacitor in terms of a Feynman diagram, it could be represented as a process that involves a mutual exchange of a virtual photon and a virtual Z^0 in parallel.** In this way the process may increase the effective total charge of the receptor plate of the capacitor (via the exchange of displacement of charge), while the simultaneous involvement of a photon γ and a Z^0 yields a zero exchange in energy in terms of matter.

- A relatively similar process seems to be taking place during a change in particles' generation. Here, a different orthogonality exists, which has to do with the exchange between the parameters of space and time. This applies as an effective change in the "gravitational" energy state of a particle (instead of a change in the "electronic" energy state). As per this orthogonality, **this process involves the simultaneous involvement of a W^+ and a W^- boson, instead of the γ and a Z^0. We may actually consider that here the W^+ and W^- bosons have the role of "a graviton and a negative graviton"**.

- In the case of quarks, the generation-change affects both the positive-matter as well as the negative-matter portion of quarks. This raises the inertial impedance coordinates with respect to the Cabibbo angles involved, where that increases the effective total mass (actually inertial impedance) of the higher generation particle. At the same time, the simultaneous use of a W^+ and a W^- yields a zero change in charge.

11
Notes on Space and Time, TOE foundation

In relation to all of the above, let us consider how the cosmos seems to function, staring from absolute scratch. Consider vacant universe, with no matter, no charge, nothing in it. Space (of any degree of stretching) and time (of any degree of fastness in passing) may exist as orthogonal entities to each other, meaning that either one is imaginary with respect to the other. That may also be taken to imply that each one may have its negative-reciprocal to counterbalance the other in terms of energy, since for imaginary entities the projection toward the real world has to involve the relation $i=-1/i$. as previously explained. This does not simply mean that one has positive energy and the other has negative energy so they sum up to zero, as it additionally requires that either one is imaginary with respect to the other (a sort of "inside-out of the other", simplistically speaking). Alongside that concept, matter has been described to refer to an oscillation in time (and the gravitational field of a particle to correspond to a counterbalancing curvature in the metric of time), and charge has been described to refer to an oscillation in negative-reciprocal space (and the electric field to co-exist as a counterbalancing curvature in the metric of negative-reciprocal space); not forgetting that from the point of reference of the electromagnetic interaction, that concerns curvature of plain "real" space. This allows for the wavinesses in these two entities to co-exist without need of some external energy supply, since the energy of both together evens-out to null. In that sense their co-existence is "allowed" by nature, in fact it appears to also be "required", for reasons of energy balancing.

Space and Time together host oscillations of various kinds, provided that the total energy sum over the whole of the cosmos is maintained at zero. In fact, any waviness in either of the two metrics comes along with a balanced reaction in the other metric. For instance, an oscillation in an electric field (whose potential concerns the curvature in the metric of negative-reciprocal space metric) comes along with an oscillation in magnetic field (whose potential was described to concern curvature in the metric of negative-reciprocal time). More extensively, alternations in Space and Time may include:

- A waviness deployed over space, for example differentials in the metric of time (which constitute the gravitational potential as per the depiction in figure 1-1) or/and the metric of space (relating to an anti-electric potential, or/and to displacement gravitomagnetic potential),
- Waviness in the negative reciprocals of the metrics of space (relating to the electric potential) and of time (relating to the magnetic potential),
- Oscillations of time and space themselves, relating to the essence of fermions' matter and charge correspondingly,
- Oscillations deploying in space (like in forming elementary particles) or deploying in time (like in electric circuit oscillations).
- Generally, we should consider that curvature of all kinds (of positive or negative time, or of positive or negative space, or their negative-reciprocal counterparts) refer to potentials, and either of those yields a corresponding force (when applied on matter, or charge, or on negative matter, or anticharge).
- Forces are counterbalanced by reaction phase shifts, in time or in space (like capacitance, inductance, mechano-capacitance, mechano-inductance), which generate the corresponding reaction potentials.
- Reaction shifts may also oscillate and interact in non-unitary space and time in the form of particles (as energy in exchanged in quantized lumps in relation to the Fourier transformation), whose field components engage through cosine and sine projections, exhibiting the characteristics of bosons.
- Due to the phase shift between the metrics of space and time (per the Weinberg angle), the above interactions may also take place fractionally, and in such way address the essense of quarks, and the strong interaction.
- All the above may apply respectively for matter, antimatter, negative matter, negative antimatter, and their electric counterparts of positive and negative charge and anticharge, with respect to the sign of the parameters of space and time that prevail at each cosmic region shown in figure 1-1.

Even though space at the cosmic scale is imagined to have endless extend, the space metric undergoes a wave-like curvature in density, as well as loop-ness, as schematically depicted in figure 1-1. So, as we move radially away from our cosmic spot, space curvature (density) may stretch up to a certain range, but after some point it starts to contract continuously until reaching a nodal point (like point d in the waviness illustrated in figure 1-1) where it turns negative (imaginary with respect to real-world point of reference), and it eventually re-attaches at point b. So, in this respect space is not infinite, it is loop-like. In fact, almost everything in nature turns out to be loop-like. What we view as infinity (referring to the node at point d) applies as a nodal surface (or, a sort of spread-out pole), in a similar sense to as point b also applies as a nodal surface (or, a sort of spread-out pole). It should not be forgotten, however, that figure 1-1 only provides a simplified picture, while the actual universe is rather more like a "sea of waves".

It similarly holds with time. We are accustomed to consider time as some sort of entity that progresses endlessly, assuming that the universe supposedly started through a big bang and now keeps expanding). But this concept doesn't appear to be correct, as time follows a loop-like waviness as well. At the cosmic scale it follows a radial-like waviness. The further away from a cosmic spot like ours, time may move faster and faster, but after some point it turns to move slower and slower until it eventually goes through a node (as in point d' of figure 1-1) where it turns negative, and after a portion which is imaginary with respect to the real world, it eventually re-attaches to real positive time at point b'. So, in this respect time is also behaving loop-like, with regions of negative direction of time simply constituting habitat for antimatter particles (and regions of negative space curvature constituting ordinary habitat for particles of negative matter respectively, like netrinos and antineutrinos). Furthermore, the waviness of time and space are not stationary, they may change value in relation to astrophysical processes, while the values of their metric in each case co-engage in determining the fastness of development and/or repetition of celestial effects (like the birth and death of stars). And along with that, the values of the corresponding metrics may engage in detenmining the gradual balancing of the life span of different species through chemical and biological interactions (while local waviness evolves).

While the above describes the loop-liness of time and space at the cosmic scale, a complementary loop-like waviness at the microscopic scale refers to energy states of fermions. In that sense, such loop-liness at various levels allows for a string-theory-like approximation to nature's interactions, without need for (counterintuitive) additional dimensions. The loop-like waviness depicted in figure 1-1 may explain the expansion of the universe without need for dark energy, there is no need for singularities, neither a big bang should have taken place in the first place. Instead, there are ever-going oscillations of the aforementioned forms, with the cosmos evolving as a "sea of waviness" of space and time, where events are happening based on causality, without the existence of "parallel universes" (in the sense of expanded causality). In terms of extra dimensions, the present theory has addressed the existence of 3 dimensions of time (instead of 1 only) alongside the 3 dimensions of space. This allows to link effects of mechanics to the gravitational interaction, and through this way reveal that the gravitational and the electromagnetic interactions are described by symmetric laws, subject to a swapping between parameters of space and time. This, therefore, bridges the gravitational interaction with the existing Standard Model. The new theory additionally introduces negative-energy field variants, namely the anti-electric, anti-magnetic, negative-gravitational, and negative gravitomagnetic fields, which typically engage in effects of reaction (like the inertial or the electromotive) through displacement-field variants. The negative-energy displacement fields also appear to accommodate the weak interaction.

An additional concept introduced, concerns the existence of phase shifts in the time domain and the space domain. The former has been accounted for (without so far realizing it) within a permittivity or permeability terms, while the latter concerns their mechanical counterparts. This allows to parallelize inductive/capacitive-like effects of mechanics to those of electronics, for example it reveals a deep symmetry between the effect of barrier penetration in mechanics, and the the crossing of charge between the plates of a capacitor in electronics. Or, it allows to explain the abundance of matter over antimatter. Furthermore, the phase shifts are explained to ride a loop-like trace, and that allows for the results of the displacement to take place concurrently (therefore instantaneously) over the whole loop (just like when turning a bicycle wheel, all elements of the wheel displace concurrently). This feat allows to explain the intantenuity behind the effect of entanglement, which today's theory treats this effect by conjecture as an effect of quantum gravity, while the present model allows to approach in classical physics terms via phase shifts. Entangled particles therefore "connect" to each other much like a capacitor and an inductor are "connected" over a circuit loop, so any shift in current affects them both instantaneously. Entanglement is currently used to communicate bits of information in quantum computing, but the new model allows to apply phase shifts in an alternating way and in this way convey ultrafast signals, providing new means for extraterrestrial communication due to transmission fastness. A corresponding method were conceived by the author, and the understanding of its particulars was actually what triggered the present research. That method (which lies beyond the purpose of this book) opens a large window for technological innovation in wireless energy transfer, remote propulsion, ultrafast communication, and energy exploitation over interdisciplinary areas of science, which constitute the scientific focus of the author. The associated theory, in the other hand, appears to restrict from chances of "time travel", by addressing issues of particle disintegration.

All the above seem to drastically clear up the puzzle of how the laws of nature bridge together, in forming a democratic place called Cosmos, where numerous interactions gear up to each other in a beautiful balance, so as to maintain that the total energy everywhere around sums up to null. As such, action and reaction beautifully intertwine in numerous ways that bring up the elegant nature we are a part of.

<u>Epilogue</u>

"An Open Call to the Scientific Community"

Thank you for reading this book. I hope that it has fulfilled its goal to shed light on certain long-standing questions of physics and science, and to provide insightful suggestions toward certain other.

A phenomenological approach has often been used, relying on symmetries traced among different parts of physics that were previously considered independent. A large emphasis has been given on the intuitive aspect.

It has been a great challenge to accommodate concepts and bridge among existing theories, with the largest difficulty being that conventional units of measurement are named after people, where that naming wouldn't reveal their deeper traits.

Due to the large number of unknowns that were in need to be matched and bridged, the work has been carried out through a very large number of iterations, meticulous work, gradual improvement, consideration and re-consideration of alternatives, search for extrapolations, and repetitive fine-tunings.

Other than being very timely, the process of developing and continuously improving concepts seems almost never ending. Therefore, I decided to publish this book to reveal the key concepts and ideas so far, as an open call to the scientific community to look into, and constructively improve and build up further upon the foundation hereby provided.

Among theoretical grounds introduced, one of particular significance concerns a new way for reading and understanding equations' imaginary terms, which also extends to reflect on quantum nature. That provides access to almost "another half" of physics, which also addresses the superluminal, opening up a very large new window for technological innovation.

In parallel to the development of the concepts provided in this book, new grounds have been invented by the author (not included in this book) which played a scientifically stirring role over this work. That concerns relevant technology which may find specific uses in domains like remote propulsion by electronic means, wireless transfer of energy, control over resonant optoelectronic applications with prospective uses in areas including quantum computing, and a novel modulation-demodulation technique for ultrafast communications through the involvement of negative energy, the latter seemingly providing the only reasonably fast way for communication with extraterrestrial intelligence.

The prospects for the future seem to be mostly promising.

Christos Tsikoudas

References

[1] "Black Holes and the Centrifugal Force Paradox".
Marek Artur Abramowicz, Scientific American, March 1993.
https://www.jstor.org/stable/24941405

[2] "The Theory of Positrons", R. P. Feynman,
Phys. Rev. 76, 749, September 1949
Doi: 10.1103/PhysRev.76.749

[3] "The Feynman-Stueckelberg picture", section 1.4,
An Introduction to QED & QCD, F. Hautmann", September 2010
https://www-thphys.physics.ox.ac.uk/people/FrancescoHautmann/
Ralhep/ralss10_p.pdf

[4] "Large-scale superluminal motion in the quasar 3C273",
Davis, R. J.; Unwin, S. C.; Muxlow, T. W. B., Nature. 354 (6352), 1991
Doi: 10.1038/354374a0

[5] "A superluminal source in the Galaxy",
Mirabel, I.F.; Rodriguez, L.F.. Nature. 371 (6492): 46-48., 1994
Doi:10.1038/371046a0

[6] "Sign of the Gravitational Mass of a Positron",
L. I. Schiff, Phys. Rev. Lett. 1, 254-255, October 1958.

[7] Bheghella-Bartoli S, Bhujbal P M, and Nas A,
"Confirmation of antimatter detection via Santilli telescope with concave
lenses", Am. J. Modern Phys., 4(3), pp. 5-11, 2014.
https://studylib.net/doc/7053859/confirmation-of-antimatter-detection-via-
santilli-telesco...

[8] Santilli R M, "Apparent detection of antimatter galaxies via a refractive
telescope with concave lenses." Clifford Analysis, Clifford Algebras and
their Applications 3.1, 2014.
https://www.prweb.com/releases/2014/02/prweb11589410.htm

[9] "Gravitational repulsion and Dirac antimatter". International Journal of
Theoretical Physics. 35 (3): 605–631. Kowitt, M., 1996.
Doi: 10.1007/BF02082828

[10] "Considerations about the apparent superluminal expansions observed in
astrophysics", Recami, Erasmo, Il Nuovo Cimento. 93, April 1986.
Doi: 10.1007/BF02722327

[11] "Superluminal motion of a relativistic jet in the neutron-star merger
GW170817", K. P. Mooley, A. T. Deller, O. Gottlieb, E. Nakar, G.
Hallinan, S. Bourke, D. A. Frail, A. Horesh, A. Corsi & K. Hotokezaka
Nature, Vol. 561, Issue 7723, p. 355-359, September 2018.
Doi: 10.1038/s41586-018-0486-3

[12] "The speed of clocks in a gravitational field",
Feynman lectures on Physics, vol. II, ch. 42-6, 1964.
http://www.feynmanlectures.caltech.edu/II_42.html

[13] "Free-space measurement of complex permittivity and complex
permeability of magnetic materials at microwave frequencies".
D.K. Ghodgaonkar, V.V. Varadan, IEEE, Vol.39, no.2, April 1990.
Doi: 10.1109/19.52520

[14] "The relativity of magnetic and electric fields",
Feynman lectures on Physics, vol. II, ch. 13-6, 1964.
https://www.feynmanlectures.caltech.edu/II_13.html

[15] "Dark matter: A primer", Advances in Astronomy, pp. 1-22,
Garrett K. and Duda G., 2011.
Doi: 10.1155/2011/968283

[16] "The extended rotation curve and the dark matter halo of M33",
Monthly Notices of the Royal Astronomical Society 311.2, pp. 441-447
Corbelli E. and Salucci P., 2000.
Doi: 10.1046/j.1365-8711.2000.03075.x

[17] "A modification of the Newtonian dynamics as a possible alternative to
the hidden mass hypothesis", Astronomical Journal, 270, pp. 365-370.
Milgrom M., 1983.
https://adsabs.harvard.edu/full/1983ApJ...270..365M

[18] "A review of astrophysical jets",
Acta Polytechnica CTU Proceedings 1.1, 259-264
Beall, James H., 2014.
Doi: 10.14311/APP.2014.01.0259

[19] "A rapidly changing jet orientation in the stellar-mass back-hole system
V404 Cygni", Nature 569.7756, pp. 374-377,
J.C.A Miller-Jones at al., 2019.
https://www.nature.com/articles/s41586-019-1152-0

[20] "Gravitomagnetic field and Penrose scattering processes", Annals of the
New York Academy of Sciences, 1045.1, pp. 232-245
R.K. Williams, 2005.
https://www.researchgate.net/profile/Reva-
Williams/publication/7764126_Gravitomagnetic_Field_and_Penrose_Scattering
_Processes/links/5a0cbdaea6fdcc39e9bfb369/Gravitomagnetic-Field-and-
Penrose-Scattering-Processes.pdf

[21] "Quantum Physics of atoms, molecules, solids, nuclei, and particles",
second edition, Robert Eisberg & Robert Resnick, Sec.5-2., 1985.
https://www.academia.edu/34441165/Eisberg_R_and_R_Resnick_
Quantum_Physics_Of_Atoms_Molecules_Solids_Nuclei_And_Particles